NEUROMETHODS

Series Editor
Wolfgang Walz
University of Saskatchewan
Saskatoon, SK, Canada

For further volumes:
http://www.springer.com/series/7657

Lipidomics

Edited by

Paul Wood

*Metabolomics Unit, Lincoln Memorial University,
Harrogate, TN, USA*

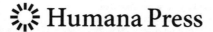 Humana Press

Editor
Paul Wood
Metabolomics Unit
Lincoln Memorial University
Harrogate, TN, USA

ISSN 0893-2336   ISSN 1940-6045   (electronic)
Neuromethods
ISBN 978-1-4939-8347-6   ISBN 978-1-4939-6946-3   (eBook)
DOI 10.1007/978-1-4939-6946-3

Printed on acid-free paper

This Humana Press imprint is published by Springer Nature
The registered company is Springer Science+Business Media LLC
The registered company address is: 233 Spring Street, New York, NY 10013, U.S.A.

Preface to the Series

Experimental life sciences have two basic foundations: concepts and tools. The *Neuromethods* series focuses on the tools and techniques unique to the investigation of the nervous system and excitable cells. It will not, however, shortchange the concept side of things as care has been taken to integrate these tools within the context of the concepts and questions under investigation. In this way, the series is unique in that it not only collects protocols but also includes theoretical background information and critiques which led to the methods and their development. Thus it gives the reader a better understanding of the origin of the techniques and their potential future development. The *Neuromethods* publishing program strikes a balance between recent and exciting developments like those concerning new animal models of disease, imaging, in vivo methods, and more established techniques, including, for example, immunocytochemistry and electrophysiological technologies. New trainees in neurosciences still need a sound footing in these older methods in order to apply a critical approach to their results.

Under the guidance of its founders, Alan Boulton and Glen Baker, the *Neuromethods* series has been a success since its first volume published through Humana Press in 1985. The series continues to flourish through many changes over the years. It is now published under the umbrella of Springer Protocols. While methods involving brain research have changed a lot since the series started, the publishing environment and technology have changed even more radically. Neuromethods has the distinct layout and style of the Springer Protocols program, designed specifically for readability and ease of reference in a laboratory setting.

The careful application of methods is potentially the most important step in the process of scientific inquiry. In the past, new methodologies led the way in developing new disciplines in the biological and medical sciences. For example, Physiology emerged out of Anatomy in the nineteenth century by harnessing new methods based on the newly discovered phenomenon of electricity. Nowadays, the relationships between disciplines and methods are more complex. Methods are now widely shared between disciplines and research areas. New developments in electronic publishing make it possible for scientists that encounter new methods to quickly find sources of information electronically. The design of individual volumes and chapters in this series takes this new access technology into account. Springer Protocols makes it possible to download single protocols separately. In addition, Springer makes its print-on-demand technology available globally. A print copy can therefore be acquired quickly and for a competitive price anywhere in the world.

Saskatoon, Canada   *Wolfgang Walz*

Preface

The "omics" technologies have significantly advanced our ability to sample a greater diversity and large sample sets of biological molecules with broad-based analytical platforms. Metabolomics is one such technology. However, metabolomics analyses are limited by the lability of a large proportion of the biomolecular pool being sampled. Lipidomics is a specialized subfield of metabolomics, with research in this area expanding at a rapid rate. This includes basic and clinical research efforts. With regard to clinical studies, a large number of plasma lipids are significantly more stable than the majority of metabolites monitored in metabolomics studies. As a result, there are a number of potential lipid biomarkers that are under intense study in the fields of oncology, infectious diseases, neurology, and psychiatry. These are biomarkers of disease, of disease progression, and of disease response to therapy. In addition, biomarkers offer the potential to stratify patient subgroups and to define new molecular targets for therapeutic intervention.

Lipids are multifunctional in that they serve as energy stores, as structural biomolecules in membranes, and in signal transduction pathways. In addition there are highly specialized lipids that serve unique biological functions. For example plasmalogens are involved in membrane fusion, seminolipids in sperm function, and (o-acyl)-ω-hydroxy-fatty acids (OAHFA) as surfactants in tears, amniotic fluid, and sperm. Electrospray ionization high-resolution mass spectrometry (ESI-HR MS) is capable of sampling across the broad domain of these lipids utilizing direct-flow analyses and introduction of samples by chromatographic methods (HPTLC, uHPLC, SFC, CZE). Tandem mass spectrometry is also essential for validating the structures of monitored lipids. Several aspects of lipidomics analyses are worthy of further discussion. First, oxidative stress that is associated with many disease sates can be monitored by quantitation of oxidized lipids. Second, lipid dynamics, with regard to signal transduction pathways, can be evaluated by characterization of the fatty acid substitutions at sn-2 of glycerophospholipids. For example, if excessive arachidonic release occurs as a result of altered signal transduction, new fatty acid substitutions at sn-2 will be monitored as a result of the tight coupling of deacylation-reacylation.

Within this volume are 16 detailed methods chapters. In addition, we have included two important review chapters. First, the excellent overview of lipidomics strategies by Dr. Griffiths puts the methods chapters in perspective. Similarly, the review of FT-ICR HR-MS by Dr. Emmett is essential since HR-MS for lipidomics is the new standard. Low resolution mass spectrometry with triple quadrupole instruments has generated a large amount of historical literature that unfortunately has a number of lipid miss-assignments due to mass errors of 50–300 ppm. FT-ICR provides the highest resolution available today and establishes the gold standard. The orbitrap mass spectrometer has resolution close to that of FT-ICR and a number of chapters in this volume discuss the use of this simpler (maintenance wise) mass spectrometer.

Our volume spans: (1) The analyses of specific lipid classes (endocannabinoids, fatty aldehydes, phospholipids, and oxidized fatty acids and phospholipids); (2) The lipidomics analyses of cells (lymphocytes and oocytes); (3) Direct-flow and chromatographic methods (SFC, UHPLC, HPTLC, ion-mobility); (4) Derivatization methods for lipids (amines, fatty aldehydes, and ketones); (5) TOF-SIMS imaging of lipids; and (6) Characterization of lipid transfer proteins. These chapters put individual lipidomics analyses in perspective in addition to supplying the critical experimental details required to conduct repeatable experiments. With the increasing dominance of the orbitrap in lipidomics studies, representative tandem mass spectra of a large number of lipid classes obtained with the orbitrap have been included as a valuable reference source via on-line Electronic Supplementary Material (Chap. 2).

In summary, lipidomics is a fast-paced research field in which HR-MS and mass spectrometric lipid imaging are advancing our knowledge of lipid dynamics, cellular ultrastructure of lipid pools, and lipid alterations in disease. An exciting decade of research is ahead for this field.

Harrogate, TN, USA   *Paul Wood*

Contents

Contributors

GIUSEPPE ASTARITA • *Waters Corporation, Milford, MA, USA; Department of Biochemistry and Molecular and Cellular Biology, Georgetown University, Washington, DC, USA*

TAKESHI BAMBA • *Department of Biotechnology, Graduate School of Engineering, Osaka University, Suita, Osaka, Japan; Division of Metabolomics, Medical Institute of Bioregulation, Kyushu University, Fukuoka, Japan*

VÁCLAV MATĚJ BIERHANZL • *Faculty of Science, Department of Analytical Chemistry, Charles University in Prague, Prague, Czech Republic*

MESUT BILGIN • *MPI of Molecular Cell Biology and Genetics, Dresden, Germany; Unit for Cell Death and Metabolism, Center for Autophagy, Recycling and Disease, Danish Cancer Society Research Center, Copenhagen, Denmark*

RADOMÍR ČABALA • *Faculty of Science, Department of Genetics and Microbiology, Charles University in Prague, Prague, Czech Republic*

R. GRAHAM COOKS • *Department of Chemistry and Center for Analytical Instrumentation Development, Purdue University, West Lafayette, IN, USA*

M. ROSÁRIO M. DOMINGUES • *Department of Chemistry and QOPNA, Mass Spectrometry Centre, University of Aveiro, Campus Santiago, Aveiro, Portugal*

PEDRO DOMINGUES • *Department of Chemistry and QOPNA, Mass Spectrometry Centre, University of Aveiro, Aveiro, Portugal*

MARK R. EMMETT • *Department of Biochemistry and Molecular Biology, University of Texas Medical Branch-Galveston (UTMB), Galveston, TX, USA; Department of Pharmacology and Toxicology, University of Texas Medical Branch-Galveston (UTMB), Galveston, TX, USA; Department of Radiation Oncology, University of Texas Medical Branch-Galveston (UTMB), Galveston, TX, USA; Mitchell Center for Neurodegenerative Diseases, University of Texas Medical Branch-Galveston (UTMB), Galveston, TX, USA; UTMB Cancer Center, University of Texas Medical Branch-Galveston (UTMB), Galveston, TX, USA*

CHRISTINA R. FERREIRA • *Department of Chemistry and Center for Analytical Instrumentation Development, Purdue University, West Lafayette, IN, USA*

TIMOTHY J. GARRETT • *Department of Pathology, Immunology and Laboratory Medicine, College of Medicine, University of Florida, Gainesville, FL, USA*

WILLIAM J. GRIFFITHS • *Swansea University Medical School, Singleton Park, Swansea UK*

ERICH GULBINS • *Department of Molecular Biology, University of Duisburg-Essen, Essen, Germany*

KENTARO HANADA • *Department of Biochemistry & Cell Biology, National Institute of Infectious Diseases, Tokyo, Japan*

ALAN K. JARMUSCH • *Collaborative Mass Spectrometry Innovation Center and Skaggs School of Pharmacy and Pharmaceutical Sciences, University of California, La Jolla, CA, USA*

ALEXANDRA C. KENDALL • *Manchester Pharmacy School, Faculty of Medical and Human Sciences, The University of Manchester, Manchester, UK*

BURKHARD KLEUSER • *Department of Nutritional Toxicology, Institute of Nutritional Science, University of Potsdam, Nuthetal, Germany*

TATIANA KONDAKOVA • *Department of Microbiology, University of Illinois at Urbana-Champaign, Urbana, IL, USA*

CHERYL F. LICHTI • *Department of Pharmacology and Toxicology, University of Texas Medical Branch-Galveston (UTMB), Galveston, TX, USA; Mitchell Center for Neurodegenerative Diseases, University of Texas Medical Branch-Galveston (UTMB), Galveston, TX, USA*

NADINE MERLET MACHOUR • *UMR 6014 COBRA, Normandy Univ., Univ. Rouen, Evreux, France*

ELISABETE MACIEL • *Department of Chemistry and QOPNA, Mass Spectrometry Centre, University of Aveiro, Aveiro, Portugal; Department of Biology and CESAM, University of Aveiro, Campus Universitário de Santiago, Aveiro, Portugal*

TÂNIA MELO • *Department of Chemistry and QOPNA, Mass Spectrometry Centre, University of Aveiro, Aveiro, Portugal*

CORINNA NEUBER • *Department of Nutritional Toxicology, Institute of Nutritional Science, University of Potsdam, Nuthetal, Germany*

ANNA NICOLAOU • *Manchester Pharmacy School, Faculty of Medical and Human Sciences, The University of Manchester, Manchester, UK*

GIUSEPPE PAGLIA • *European Academy of Bolzano/Bozen, Center for Biomedicine, Bolzano, Italy*

VALENTINA PIRRO • *Department of Chemistry and Center for Analytical Instrumentation Development, Purdue University, West Lafayette, IN, USA*

CÉCILE DUCLAIROIR POC • *Laboratory of Microbiology Signals and Microenvironment (LMSM) EA4312, Normandy Univ., Univ. Rouen, Evreux, France*

ANA REIS • *Department of Chemistry and QOPNA, Mass Spectrometry Centre, University of Aveiro, Aveiro, Portugal*

MARTINA RIESOVÁ • *Faculty of Science, Department of Analytical Chemistry, Charles University in Prague, Prague, Czech Republic; Institute of Forensic Medicine and Toxicology, General University Hospital in Prague, Prague, Czech Republic*

FABIAN SCHUMACHER • *Department of Nutritional Toxicology, Institute of Nutritional Science, University of Potsdam, Nuthetal, Germany; Department of Molecular Biology, University of Duisburg-Essen, Essen, Germany*

GABRIELA SEYDLOVÁ • *Faculty of Science, Department of Physical Chemistry, Charles University in Prague, Prague, Czech Republic*

ANDREJ SHEVCHENKO • *MPI of Molecular Cell Biology and Genetics, Dresden, Germany*

BINDESH SHRESTHA • *Waters Corporation, Milford, MA, USA*

TOSHIHIKO SUGIKI • *Institute for Protein Research, Osaka University, Suita, Osaka, Japan*

CANDICE Z. ULMER • *Department of Chemistry, University of Florida, Gainesville, FL, USA*

YUQIN WANG • *Swansea University Medical School, Singleton Park, Swansea, UK*

NORELLE C. WILDBURGER • *Department of Neurology, Washington University School of Medicine, St. Louis, MO, USA*

PAUL WOOD • *Metabolomics Unit, Lincoln Memorial University, Harrogate, TN, USA*

TAKAYUKI YAMADA • *Department of Biotechnology, Graduate School of Engineering, Osaka University, Suita, Osaka, Japan*

RICHARD A. YOST • *Department of Chemistry, University of Florida, Gainesville, FL, USA; Department of Pathology, Immunology and Laboratory Medicine, College of Medicine, University of Florida, Gainesville, FL, USA*

Chapter 1

Introduction and Overview of Lipidomic Strategies

William J. Griffiths and Yuqin Wang

Abstract

While analysis of the genome or proteome can be predictive of the fate of an organism, its metabolome is informative of the outcome of events on an organism. The metabolome can thus be regarded as closer to the actual phenotype than either the genome or proteome. Lipids represent a major component of the metabolome and their hydrophobic and amphipathic nature dictates their separate analysis from more water-soluble metabolites; hence, the discipline of lipidomics has emerged. In this short chapter, we will highlight the different strategies in lipidomics which have now reached maturity, i.e., shotgun and chromatography-based mass spectrometry approaches. We will discuss some of the newer technologies coming to the forefront e.g., the use of derivatization chemistry, and comment on exciting developments now being made in the lipidomic field, e.g., surface analysis and lipid imaging. Finally, we will comment on some of the dangers encountered using an "omics" approach in biochemical analysis.

Key words Lipidomics strategies, Analytical approaches, Lipid imaging

1 Introduction

Lipidomics can be defined as the quantitative identification of all lipids in an organism, tissue, fluid, or cell type. Lipids themselves are defined as hydrophobic or amphipathic small molecules that may originate entirely or in part by carbanion-based condensations of thioesters (e.g., fatty acyls) and/or by carbocation-based condensations of isoprene units (e.g., prenols, sterols) [1]. An older less exact definition of lipids, often found in general biochemistry texts, is naturally occurring compounds readily soluble in organic solvents. Lipids represent a major part of the metabolome [2]; however, the insolubility of many lipids in aqueous solutions dictates their separate analysis from the more water-soluble components of the metabolome. Although nuclear magnetic resonance (NMR) has been used wildly in metabolomic studies [3], mass spectrometry (MS) is the dominant technology in lipidomics. Historically, gas chromatography (GC)–MS was a major tool for lipid profiling studies [4, 5], but today this technology has been overtaken by the interfacing of desorption ionization techniques of

Paul Wood (ed.), *Lipidomics*, Neuromethods, vol. 125,
DOI 10.1007/978-1-4939-6946-3_1, © Springer Science+Business Media LLC 2017

atmospheric pressure ionization (API) and matrix-assisted laser desorption/ionization (MALDI) with MS. API, usually electrospray ionization (ESI) or atmospheric pressure chemical ionization (APCI), is often linked to a liquid chromatography (LC) system providing separation prior to MS analysis. MALDI is mostly performed under vacuum and is not usually linked to LC separation, although MALDI can be performed from thin-layer chromatography (TLC) plates. MALDI –MS imaging (MSI) is becoming an ever more adopted technology in lipidomics providing spatial information to lipid classes in tissues [6].

MS gives molecular weight information via measurement of the mass-to-charge ratio (m/z) of ionized species and many modern MS instruments are capable of providing mass accuracy at a level of 0.001–0.002 m/z. This degree of mass accuracy allows m/z searches against, e.g., the Lipid Maps database of compounds http://www.lipidmaps.org/ or through various commercial or open-source software, e.g., XCMS, MZmine to identify ionized molecules [7–9]. Structural information is provided by performing tandem mass spectrometry (MS/MS) or multistage fragmentation (MS^n) experiments. Databases for identifications from MS/MS data are in existence, e.g., LipidBlast, MS-DIAL [10, 11], while the Lipid Maps website contains MS/MS spectra for over 500 lipids. While reliable annotations can be made from MS and MS/MS data, further confidence of identification can be made by combining chromatographic separation as in GC-MS and LC-MS. Definitive identification requires comparison of retention time, mass, and MS/MS fragmentation with an authentic standard. A European group have provided a useful nomenclature to annotate lipids from MS data indicating the extent of structural certainty [12].

MS not only provides information for lipid identification, it can also be used for lipid quantification. Ideally, quantification is performed with the use of isotope-labeled standards added during sample preparation, or if these are not available via structurally related analogues. Avanti Polar Lipids http://avantilipids.com/ provide a large number of high-quality authentic standards imperative for accurate quantification.

There are essentially two strategies in lipidomics analysis. There is the global "shotgun" approach developed by Han and colleagues in the USA [13] and by Shevchenko and coworkers in Europe [14], where the aim is to identify/quantify a maximum number of lipid compounds with a minimum amount of sample preparation. This was initially performed with direct-infusion ESI, but now more commonly chip-based nano-ESI is employed. The main disadvantage of shotgun lipidomics is its bias toward more abundant and readily ionized lipids. This has been overcome by both Han and Schevchenko's groups by clever use of derivatization chemistry. The alternative strategy in lipidomics is the targeted approach where each class of lipids is analyzed one at a time (Fig. 1). This approach has been impressively exploited by the Lipid Maps consortium in the USA where different

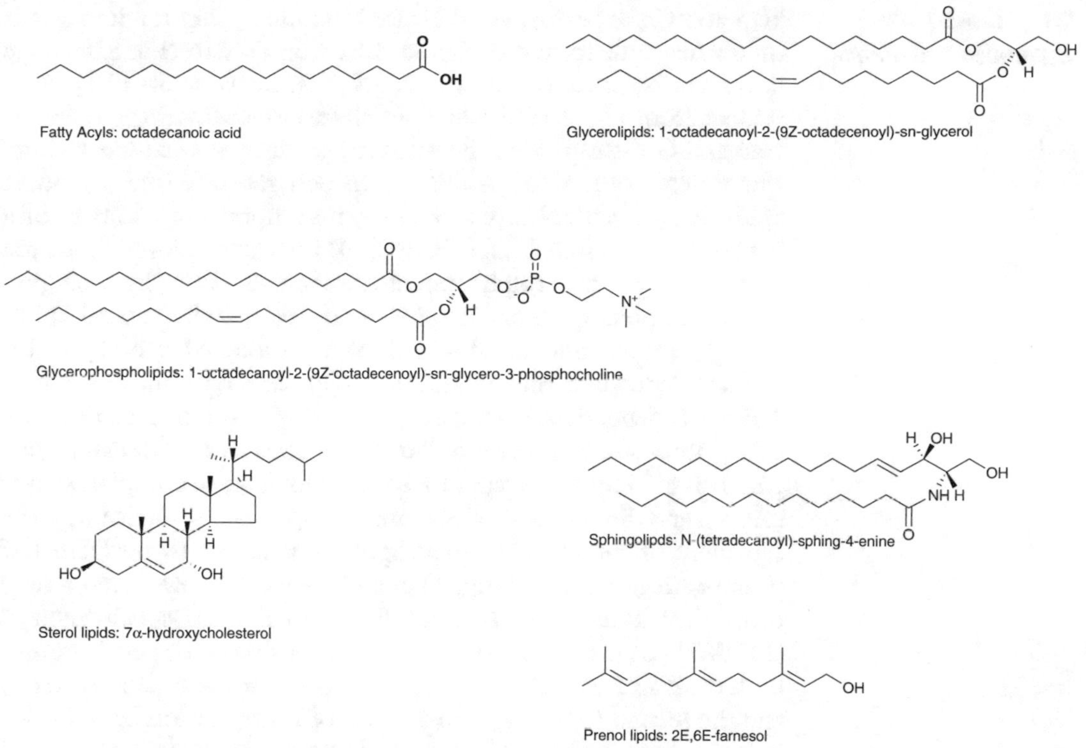

Fatty Acyls: octadecanoic acid

Glycerolipids: 1-octadecanoyl-2-(9Z-octadecenoyl)-sn-glycerol

Glycerophospholipids: 1-octadecanoyl-2-(9Z-octadecenoyl)-sn-glycero-3-phosphocholine

Sterol lipids: 7α-hydroxycholesterol

Sphingolipds: N-(tetradecanoyl)-sphing-4-enine

Prenol lipids: 2E,6E-farnesol

Fig. 1 Six classes of lipids commonly analyzed in lipidomics studies

groups have combined data for the lipidomic analysis of a pooled plasma sample, RAW 264.7 immortalized macrophage-like cell, and tissue, plasma, and urine from nonalcoholic fatty liver disease (NAFLD) includes steatosis, nonalcoholic steatohepatitis (NASH) patients [15–17]. The Lipid Maps consortium have posted their data and all protocols on their website http://www.lipidmaps.org/. The analytical methods used can also be found on this site and are published in an issue of Methods in Enzymology [18].

2   Shotgun Lipidomics

The term shotgun lipidomics was coined by Han and Gross and the methodology was excellently reviewed in their 2005 publication [13]. In the succeeding decade shotgun lipidomics has been developed further and has been exploited by many other workers, but the fundamental concepts remain the same [19]. The idea is to perform a lipid extraction, usually using chloroform and methanol [13], or alternatively methyl-tert-butyl ether (MTBE) [20], and analyze the crude extract directly by ESI-MS/MS [13]. The scanning protocols used depend on the mass spectrometer available, but the goal is always to identify and quantify as many lipid molecular species as possible.

Han and Gross performed their initial studies using tandem quadrupole instruments and designed their workflow to take advantage of the performance characteristics of these instruments [13]. As is evident from Fig. 1 different lipid classes have differing polarities. Han and Gross exploited this to perform "intra-source separation" where negative-ion ESI is utilized to preferentially ionize anionic lipids, e.g., glycerophosphates (PA), then upon the addition of a basic solution, often NH_4OH or LiOH, weakly anionic lipids are analyzed, e.g., glycerolphosphoethanolamines (PE). By changing ion-source polarity to positive-ion mode, electrically neutral lipids, e.g., glycerophosphocholines (PC) become ionized as NH_4^+ or Li^+ adducts depending on the additive used. The optimum choice of additive is dependent on the exact design of ESI interface used. By using "intra-source separation" different classes of lipids can be targeted in different scans, circumventing problems of completive ionization and limitations of dynamic range, while exploiting the capability of the tandem quadrupole instrument to perform fast positive/negative switching. Han and Gross also developed a protocol they latter named "multidimensional mass spectrometry" (MDMS) [21]. The first (x) dimension consists of ESI-MS scans in both the positive- and negative-ion modes (also called survey scans) and the second (y) dimension consists of a progression of collision-induced dissociation (CID) neutral-loss and precursor-ion MS/MS scans, again in both the positive- and negative-ion modes. Cross-peaks between the two dimensions represent fragments of a particular precursor. Quantification is performed by the addition of internal standards during extraction, a least one standard for each class of lipid. On tandem quadrupole instruments MDMS provides a sensitivity advantage over conventional product-ion scans (where the entire m/z range from the precursor ion downward is monitored), as time and sample is not wasted monitoring redundant fragment-ions. Shotgun lipidomics exploiting "intra-source separation" and MDMS works very well for glycerolipids [22], glycerophospholipids [13], and sphingolipids [23], but less well for other types of lipids that are difficult to ionize or are in low abundance. To overcome this Han and colleagues have exploited a suite of derivatization reactions [22, 24–26]. MDMS takes advantage of the ordered nature of lipid molecules (Fig. 1) which generally consist of a polar head group, a linker, and an aliphatic chain. In biology the number of different head groups is limited as is type of aliphatic chain. The nature of the lipid head group can often be established in a neutral-loss or precursor-ion scan, e.g., neutral loss of 87 Da [$NHCH(CH_2OH)CO$] is a characteristic of [M-H]$^-$ ions glycerophosphoserines (PS), a product ion scan for m/z 184 ([$H_2O_3POC_2H_4N(CH_3)_3$]$^+$) is a characteristic of choline containing lipids (PC or phosphosphingolipids), and the product-ion m/z 196 [$CH_2OCHCH_2OPO_3C_2H_4NH_2$]$^-$ is a characteristic of PE [13]. Precursor-ion scans in the negative-ion mode for [RCO_2]$^-$, or neutral-loss scans in the positive-ion mode for the loss of RCO_2H, reveal fatty acid substituents [22]. Han and colleagues are able to scan for over 30,000 potential lipid molecular species and have identified as many as 2000 from a chloroform extract [27, 28]. Shotgun lipidomics is also performed on other types of mass spectrometer. When Q-TOF type instruments are utilized following the recording of a survey scan (x-dimension), MS/MS spectra are recorded for each m/z (y-dimension). Reconstructed-ion chromatograms (RIC) can then be generated to match structurally specific product ions with their precursors [29].

2.2 Top-Down Lipidomics

Shevchenko and colleagues in Dresden have developed a shotgun lipidomics approach, called "top-down lipidomics," exploiting the attributes of high-resolution mass spectrometers [14, 30]. They have utilized Orbitrap mass spectrometers that not only give high resolving power (>100,000 full width at half maximum height) but also accurate mass measurements (<5 ppm). By accurately measuring the m/z of an ion its likely elemental composition can be determined, if not exactly, then down to only a few possibilities. This leads to lipid annotation, if not definitive identification. Identification can be made by recording an MS/MS spectrum and comparing it to an experimentally or theoretically derived spectrum in a database. Schevchenko and colleagues have developed a software to automate this process [31]. Top-down lipidomics is usually performed using automated chip-based nano-ESI infusion. Spectra can be recorded in about 1 min, allowing hundreds of samples to be analyzed in 1 day. Quantification is performed by the addition of internal standards. As with the methods developed by Han and colleagues mentioned above, "top-down lipidomics" is biased, discriminating against low abundance and poorly ionized lipids. Again, to overcome this problem Shevchenko and colleagues have made use of clever derivatization chemistry [32].

2.3 Shotgun Steroidomics

While it is fair to say that in the era of ESI the dominant methodologies have been those developed by the groups, and colleagues, of Han and Schevchenko, in the preceding era dominated by fast atom bombardment (FAB)-MS and liquid secondary ionization mass spectrometry (LSIMS) similar methods were used for metabolite profiling, particularly for steroids and bile acids. FAB-MS and/or LSIMS have been extensively used to profile steroids and bile acids in urine [5, 33–35]. Steroids are usually excreted in urine as sulfate or glucuronide conjugates making them suitable for ionization in the negative-ion mode, similarly the carboxylate group of bile acids facilitates negative-ion analysis. Bile acids may be conjugated with taurine or glycine and sulfuric, glucuronic acid, or other sugars, further encouraging negative-ion formation. For urine analysis bile acids or steroids are first extracted on a C_{18} or similar reversed-phase solid-phase extraction cartridge and the methanol or ethanol eluate is suitable for analysis. Today, ESI-MS has replaced FAB-MS and LSIMS for steroid and bile acid analysis [34, 35].

2.4 LC-Shotgun Lipidomics

While the classic shotgun lipidomics experiment involves direct infusion of a lipid extract, this can be modified by the up-stream inclusion of LC column offering some separation before ionization [36, 37]. The LC mobile phase must have a high organic content to provide for lipid solubility. The advantage provided by including a LC separation step to the shotgun format is that retention time information provides an extra dimension for analyte identification. In a 2010 study, Sandra et al. were able to detect over 1500 features (retention time, m/z pairs) from a plasma lipid extract [38]. The inclusion of an LC step does have its disadvantages. Throughput is reduced as the LC separation will have a defined run time, cross-contamination between injections is always a worry as is overloading of the column and distorting quantitative measurements. It should also be remembered the cruder the sample the more likely the column will become blocked or its performance deteriorate.

3 Targeted Lipidomics

An alternative to the shotgun lipidomic approach is provided by targeted lipidomics, where each class of lipid is analyzed one by one, exploiting extraction and MS procedures specifically designed for the target class of analyte. This approach is the one adopted by the Lipid Maps consortium in the US http://www.lipidmaps.org/. The extraction and MS procedures used to analyze each class of lipid are published in a special issue of Methods in Enzymology http://www.lipidmaps.org/resources/downloads/2007_methods_chapters.pdf [18]. The consortium have published their lipidomic analysis of the RAW 264.7 macrophage-like cell line in which they identified 229 lipid species including 163 glycerophospholipids, 48 sphingolipids, 13 sterols, and 5 prenols [39]. Following stimulation by lipopolysaccharide Kdo2-lipid A, to mimic bacterial infection, and activation of TLR4, increases in the levels of ceramides and cholesterol precursors were noted. In a similar study of the human plasma lipidome more than 500 distinct lipid species were detected and quantified including 73 glycerolipids, 160 glycerophospholipids, 204 sphingolipids, 76 eicosanoids, 31 fatty acids, 8 prenols, and 36 sterols [15].

3.1 Targeted Lipidomics Exploiting Derivatization

The idea of using derivatization chemistry to target specific functional groups and enhance the analysis of compounds possessing these groups has been exploited in MS for decades [40]. This concept is now gaining popularity in modern lipidomics. Wang et al. devised a derivatization method for the analysis of lipids with a 4-hydroxyalkenal structure by exploiting the Michael reaction with

carnosine [25]. 4-Hydroxyalkenal species are peroxidation products of polyunsaturated fatty acids and serve as "toxic second
messengers" in cellular systems. Following their derivatization,
ionization in the positive ion-mode is enhanced and derivative-
specific fragment ions facilitate precursor-ion determination [25].
Han's group also exploited derivatization chemistry for the analysis
of fatty acids. In a one-step reaction the carboxyl group of the fatty
acid is reacted with N-(4-aminomethylphenyl)pyridinium (AMPP)
in the presence of a carbodiimide catalyst [26]. The AMPP reagent
is positively charged; thus, the derivatized molecules give high ion
currents upon ESI-MS analysis. An added advantage is that MS/
MS patterns of the derivatized molecules are structurally informative. The AMPP reagent was, in fact, first used by Bollinger et al.
to aid in the ESI-MS analysis of eicosanoids [41]. Derivatization
strategies have also been applied by Reid and colleagues to intact
plasmalogen glycerophospholipids [42]. They targeted the
O-alkenyl-ether double bond within plasmenyl ethers by derivatization with iodine and methanol. The derivatized molecules show
characteristic mass shifts and give informative fragment ions upon
MS/MS. They also derivatized amine-containing plasmalogens
with $^{13}C_1$-S,S'-dimethylthiobutanoylhydroxysuccinimide ester
($^{13}C_1$-DMBNHS) reagent prior to iodine and methanol addition.
Again, derivatization provides a characteristic mass shift and informative fragment ions are forthcoming upon MS/MS.

We have developed a derivatization strategy for the analysis of
sterols and steroids. Inspired by the use of the Girard hydrazine
reagents by Shackleton et al. to enhance the ESI-MS ion-current
for steroids possessing a carbonyl group [43] and the use of micro-
chemical reactions by Brooks and colleagues [44], we have developed a technology we call enzyme-assisted derivatization for
steroid analysis (EADSA) [45–47]. Steroids, including sterols and
bile acids, possessing an oxo group are derivatized directly with the
Girard P (GP) hydrazine reagent, those with no oxo group but
with a 3β-hydroxy-5-ene or 3β-hydroxy-5α-hydrogen structure are
first treated with cholesterol oxidase to introduce a 3-oxo group
and are then derivatized with the GP reagent, while steroids with a
3α-hydroxy group are oxidized with 3α-hydroxysteroid dehydrogenase to the 3-oxo equivalents and then derivatized with the GP
reagent. Not only does GP derivatization enhance ionization in
both ESI and MALDI, it also directs fragmentation in MS/MS.
This technology is now being exploited by others [48, 49].

While most chemical derivatization reactions are performed
prior to MS analysis. Derivatization can also be exploited within
the mass spectrometer itself. Blanksby, Mitchell, and colleagues in
Australia have developed a derivatization method they call OzID in
which selected precursor ions are subjected to gas phase reaction
with ozone in a linear ion-trap [50]. The resulting CID/OzID
spectra are particularly valuable for determining sn-position and
carbon double bond positions in glycerophospholipids.

4   Surface Analysis and Mass Spectrometry Imaging

Modern mass spectrometry imaging (MSI) has been pioneered by Richard Caprioli and colleagues in Vanderbilt University [51]. His group has largely used MALDI-MSI for the imaging of proteins and peptides. In a MALDI-MSI experiment tissue surface is coated with matrix and the otherwise unadulterated sample analyzed by MALDI-MS. By moving the MALDI plate relative to the laser beam mass spectra are recorded over the tissue. Modern systems give 10–50 μm lateral resolution and maps of specific ion abundances can be generated to visualize the locations of defined molecules in a tissue slice. MALDI-MSI is particularly suitable for lipid analysis as the MALDI matrix appears to extract lipids from the cut surface. Murphy and colleagues have extensively used this technology for lipid imaging [6]. While lipid imaging works very well for glycerophospholipids and other abundant lipids, it works less well for poorly ionized lipids. To overcome this derivatization chemistry can be exploited. This has been demonstrated by Andrew and colleagues for steroid analysis in brain [52].

Surface analysis can also be achieved using ESI via the technique desorption electrospray ionization (DESI) developed by Graham Cooks and colleagues in Purdue [53]. DESI generally provides lower lateral resolution than MALDI-MSI but avoids the necessity for matrix application.

5   Conclusions

Lipidomics has progressed over the last decade to a mature technology [54]. There are still major challenges ahead, e.g., in the localization of lipids to defined spatial regions, the observation of their temporal changes, and the analysis of lipids that may be minor in most situations and only become abundant after appropriate stimulation. There are also dangers for lipid analysis using an omics approach. The availability of metabolite and lipid databases has led to a careless use of the term identification. To an analytical chemist a compound is only identified when its chemical and physical properties exactly match those of an authentic standard. When only a single property, e.g., exact mass, is matched to a structure in a database, the term annotate is more appropriate. Care must also be taken when performing quantification. The gold standard for quantification is isotope dilution MS where an isotope labeled standard is added during analyte extraction. This may not always be possible in which case quantification is best performed against structural analogues. Finally, with respect to quantification, the analyst should be aware of the precision of their method by analysis of repeated quality control samples before claiming changes in analyte concentrations are a consequence of a biological pertebation. With these caveats in mind the future is bright.

Acknowledgments

This work was supported by the UK Biotechnology and Biological Sciences Research Council (BBSRC, grant numbers BB/I001735/1 to WJG, BB/L001942/1 to YW).

References

1. Brown HA, Murphy RC (2009) Working towards an exegesis for lipids in biology. Nat Chem Biol 5(9)602–606
2. Sud M, Fahy E, Cotter D, Azam K, Vadivelu I, Burant C, Edison A, Fiehn O, Higashi R, Nair KS, Sumner S, Subramaniam S (2016) Metabolomics workbench: an international repository for metabolomics data and metadata, metabolite standards, protocols, tutorials and training, and analysis tools. Nucleic Acids Res 44(D1)D463–D470
3. Dunn WB, Broadhurst DI, Atherton HJ, Goodacre R, Griffin JL (2011) Systems level studies of mammalian metabolomes: the roles of mass spectrometry and nuclear magnetic resonance spectroscopy. Chem Soc Rev 40(1)387–426
4. Ryhage R, Stenhagen E (1960) Mass spectrometry in lipid research. J Lipid Res 1:361–390
5. Sjövall J (2004) Fifty years with bile acids and steroids in health and disease. Lipids 39(8)703–722
6. Murphy RC, Hankin JA, Barkley RM, Zemski Berry KA (2011) MALDI imaging of lipids after matrix sublimation/deposition. Biochim Biophys Acta 1811(11)970–975
7. Subramaniam S, Fahy E, Gupta S, Sud M, Byrnes RW, Cotter D, Dinasarapu AR, Maurya MR (2011) Bioinformatics and systems biology of the lipidome. Chem Rev 111(10)6452–6490
8. Gowda H, Ivanisevic J, Johnson CH, Kurczy ME, Benton HP, Rinehart D, Nguyen T, Ray J, Kuehl J, Arevalo B, Westenskow PD, Wang J, Arkin AP, Deutschbauer AM, Patti GJ, Siuzdak G (2014) Interactive XCMS Online: simplifying advanced metabolomic data processing and subsequent statistical analyses. Anal Chem 86(14)6931–6939
9. Pluskal T, Castillo S, Villar-Briones A, Oresic M (2010) MZmine 2: modular framework for processing, visualizing, and analyzing mass spectrometry-based molecular profile data. BMC Bioinformatics 11:395
10. Kind T, Liu KH, Lee do Y, DeFelice B, Meissen JK, Fiehn O (2013) LipidBlast in silico tandem mass spectrometry database for lipid identification. Nat Methods 10(8)755–758
11. Tsugawa H, Cajka T, Kind T, Ma Y, Higgins B, Ikeda K, Kanazawa M, VanderGheynst J, Fiehn O, Arita M (2015) MS-DIAL: data-independent MS/MS deconvolution for comprehensive metabolome analysis. Nat Methods 12(6)523–526
12. Liebisch G, Vizcaíno JA, Köfeler H, Trötzmüller M, Griffiths WJ, Schmitz G, Spener F, Wakelam MJ (2013) Shorthand notation for lipid structures derived from mass spectrometry. J Lipid Res 54(6)1523–1530
13. Han X, Gross RW (2005) Shotgun lipidomics: electrospray ionization mass spectrometric analysis and quantitation of cellular lipidomes directly from crude extracts of biological samples. Mass Spectrom Rev 24(3)367–412
14. Schwudke D, Hannich JT, Surendranath V, Grimard V, Moehring T, Burton L, Kurzchalia T, Shevchenko A (2007) Top-down lipidomic screens by multivariate analysis of high-resolution survey mass spectra. Anal Chem 79(11)4083–4093
15. Quehenberger O, Armando AM, Brown AH, Milne SB, Myers DS, Merrill AH, Bandyopadhyay S, Jones KN, Kelly S, Shaner RL, Sullards CM, Wang E, Murphy RC, Barkley RM, Leiker TJ, Raetz CR, Guan Z, Laird GM, Six DA, Russell DW, McDonald JG, Subramaniam S, Fahy E, Dennis EA (2010) Lipidomics reveals a remarkable diversity of lipids in human plasma. J Lipid Res 51(11)3299–3305
16. Dennis EA, Deems RA, Harkewicz R, Quehenberger O, Brown HA, Milne SB, Myers DS, Glass CK, Hardiman G, Reichart D, Merrill AH Jr, Sullards MC, Wang E, Murphy RC, Raetz CR, Garrett TA, Guan Z, Ryan AC, Russell DW, McDonald JG, Thompson BM, Shaw WA, Sud M, Zhao Y, Gupta S, Maurya MR, Fahy E, Subramaniam S (2010) A mouse macrophage lipidome. J Biol Chem 285(51)39976–39985

17. Gorden DL, Myers DS, Ivanova PT, Fahy E, Maurya MR, Gupta S, Min J, Spann NJ, McDonald JG, Kelly SL, Duan J, Sullards MC, Leiker TJ, Barkley RM, Quehenberger O, Armando AM, Milne SB, Mathews TP, Armstrong MD, Li C, Melvin WV, Clements RH, Washington MK, Mendonsa AM, Witztum JL, Guan Z, Glass CK, Murphy RC, Dennis EA, Merrill AH Jr, Russell DW, Subramaniam S, Brown HA (2015) Biomarkers of NAFLD progression: a lipidomics approach to an epidemic. J Lipid Res 56(3)722–736

18. Brown HA (2007) Lipidomics and bioactive lipids: specialized analytical methods and lipids in disease. Preface Methods Enzymol 433:XV–XVI

19. Wang M, Wang C, Han RH, Han X (2016) Novel advances in shotgun lipidomics for biology and medicine. Prog Lipid Res 61:83–108

20. Matyash V, Liebisch G, Kurzchalia TV, Shevchenko A, Schwudke D (2008) Lipid extraction by methyl-tert-butyl ether for high-throughput lipidomics. J Lipid Res 49(5)1137–1146

21. Han X, Yang K, Gross RW (2012) Multi-dimensional mass spectrometry-based shotgun lipidomics and novel strategies for lipidomic analyses. Mass Spectrom Rev 31(1)134–178

22. Gross RW, Han X (2009) Shotgun lipidomics of neutral lipids as an enabling technology for elucidation of lipid-related diseases. Am J Physiol Endocrinol Metab 297(2)E297–E303

23. Cheng H, Jiang X, Han X (2007) Alterations in lipid homeostasis of mouse dorsal root ganglia induced by apolipoprotein E deficiency: a shotgun lipidomics study. J Neurochem 101(1)57–76

24. Jiang X, Ory DS, Han X (2007) Characterization of oxysterols by electrospray ionization tandem mass spectrometry after one-step derivatization with dimethylglycine. Rapid Commun Mass Spectrom 21(2)141–152

25. Wang M, Fang H, Han X (2012) Shotgun lipidomics analysis of 4-hydroxyalkenal species directly from lipid extracts after one-step in situ derivatization. Anal Chem 84(10)4580–4586

26. Wang M, Han RH, Han X (2013) Fatty acidomics: global analysis of lipid species containing a carboxyl group with a charge-remote fragmentation-assisted approach. Anal Chem 85(19)9312–9320

27. Han X, Yang K, Gross RW (2008) Microfluidics-based electrospray ionization enhances the intrasource separation of lipid classes and extends identification of individual molecular species through multi-dimensional mass spectrometry: development of an automated high-throughput platform for shotgun lipidomics. Rapid Commun Mass Spectrom 22(13)2115–2124

28. Yang K, Cheng H, Gross RW, Han X (2009) Automated lipid identification and quantification by multidimensional mass spectrometry-based shotgun lipidomics. Anal Chem 81(11)4356–4368

29. Ejsing CS, Duchoslav E, Sampaio J, Simons K, Bonner R, Thiele C, Ekroos K, Shevchenko A (2006) Automated identification and quantification of glycerophospholipid molecular species by multiple precursor ion scanning. Anal Chem 78(17)6202–6214

30. Graessler J, Schwudke D, Schwarz PE, Herzog R, Shevchenko A, Bornstein SR (2009) Top-down lipidomics reveals ether lipid deficiency in blood plasma of hypertensive patients. PLoS One 4(7)e6261

31. Herzog R, Schwudke D, Shevchenko A (2013) LipidXplorer: software for quantitative shotgun lipidomics compatible with multiple mass spectrometry platforms. Curr Protoc Bioinformatics 43:14.12

32. Lavrynenko O, Nedielkov R, Möller HM, Shevchenko A (2013) Girard derivatization for LC-MS/MS profiling of endogenous ecdysteroids in Drosophila. J Lipid Res 54(8)2265–2272

33. Setchell KD, Heubi JE (2006) Defects in bile acid biosynthesis-diagnosis and treatment. J Pediatr Gastroenterol Nutr 43(Suppl 1) S17–S22

34. Clayton PT (2011) Disorders of bile acid synthesis. J Inherit Metab Dis 34(3)593–604

35. Shackleton CH (2012) Role of a disordered steroid metabolome in the elucidation of sterol and steroid biosynthesis. Lipids 47(1)1–12

36. Chitraju C, Trötzmüller M, Hartler J, Wolinski H, Thallinger GG, Lass A, Zechner R, Zimmermann R, Köfeler HC, Spener F (2012) Lipidomic analysis of lipid droplets from murine hepatocytes reveals distinct signatures for nutritional stress. J Lipid Res 53(10)2141–2152

37. Cífková E, Holčapek M, Lísa M, Vrána D, Melichar B, Študent V (2015) Lipidomic differentiation between human kidney tumors and surrounding normal tissues using HILIC-HPLC/ESI-MS and multivariate data analysis. J Chromatogr B Analyt Technol Biomed Life Sci 1000:14–21

38. Sandra K, Pereira Ados S, Vanhoenacker G, David F, Sandra P (2010) Comprehensive blood plasma lipidomics by liquid chromatography/quadrupole time-of-flight mass spectrometry. J Chromatogr A 1217(25)4087–4099

39. Andreyev AY, Fahy E, Guan Z, Kelly S, Li X, McDonald JG, Milne S, Myers D, Park H, Ryan A, Thompson BM, Wang E, Zhao Y, Brown HA, Merrill AH, Raetz CR, Russell DW, Subramaniam S, Dennis EA (2010) Subcellular organelle lipidomics in TLR-4-activated macrophages. J Lipid Res 51(9)2785–2797

40. Halket JM, Zaikin VG (2003) Derivatization in mass spectrometry-1. Silylation. Eur J Mass Spectrom (Chichester, Eng) 9(1)1–21

41. Bollinger JG, Thompson W, Lai Y, Oslund RC, Hallstrand TS, Sadilek M, Turecek F, Gelb MH (2010) Improved sensitivity mass spectrometric detection of eicosanoids by charge reversal derivatization. Anal Chem 82(16)6790–6796

42. Fhaner CJ, Liu S, Zhou X, Reid GE (2013) Functional group selective derivatization and gas-phase fragmentation reactions of plasmalogen glycerophospholipids. Mass Spectrom (Tokyo) 2(Spec Iss)S0015

43. Shackleton CH, Chuang H, Kim J, de la Torre X, Segura J (1997) Electrospray mass spectrometry of testosterone esters: potential for use in doping control. Steroids 62(7)523–529

44. Brooks CJ, Cole WJ, Lawrie TD, MacLachlan J, Borthwick JH, Barrett GM (1983) Selective reactions in the analytical characterisation of steroids by gas chromatography-mass spectrometry. J Steroid Biochem 19(1A)189–201

45. Karu K, Hornshaw M, Woffendin G, Bodin K, Hamberg M, Alvelius G, Sjövall J, Turton J, Wang Y, Griffiths WJ (2007) Liquid chromatography-mass spectrometry utilizing multi-stage fragmentation for the identification of oxysterols. J Lipid Res 48(4)976–987

46. Griffiths WJ, Crick PJ, Wang Y, Ogundare M, Tuschl K, Morris AA, Bigger BW, Clayton PT, Wang Y (2013) Analytical strategies for characterization of oxysterol lipidomes: liver X receptor ligands in plasma. Free Radic Biol Med 59:69–84

47. Crick PJ, William Bentley T, Abdel-Khalik J, Matthews I, Clayton PT, Morris AA, Bigger BW, Zerbinati C, Tritapepe L, Iuliano L, Wang Y, Griffiths WJ (2015) Quantitative charge-tags for sterol and oxysterol analysis. Clin Chem 61(2)400–411

48. Roberg-Larsen H, Lund K, Seterdal KE, Solheim S, Vehus T, Solberg N, Krauss S, Lundanes E, Wilson SR (2016) Mass spectrometric detection of 27-hydroxycholesterol in breast cancer exosomes. J Steroid Biochem Mol Biol. pii: S0960–0760(16)30020–6

49. Soroosh P, Wu J, Xue X, Song J, Sutton SW, Sablad M, Yu J, Nelen MI, Liu X, Castro G, Luna R, Crawford S, Banie H, Dandridge RA, Deng X, Bittner A, Kuei C, Tootoonchi M, Rozenkrants N, Herman K, Gao J, Yang XV, Sachen K, Ngo K, Fung-Leung WP, Nguyen S, de Leon-Tabaldo A, Blevitt J, Zhang Y, Cummings MD, Rao T, Mani NS, Liu C, McKinnon M, Milla ME, Fourie AM, Sun S (2014) Oxysterols are agonist ligands of RORγt and drive Th17 cell differentiation. Proc Natl Acad Sci U S A 111(33)12163–12168

50. Pham HT, Maccarone AT, Thomas MC, Campbell JL, Mitchell TW, Blanksby SJ (2014) Structural characterization of glycerophospholipids by combinations of ozone- and collision-induced dissociation mass spectrometry: the next step towards "top-down" lipidomics. Analyst 139(1)204–214

51. Norris JL, Caprioli RM (2013) Analysis of tissue specimens by matrix-assisted laser desorption/ionization imaging mass spectrometry in biological and clinical research. Chem Rev 113(4)2309–2342

52. Cobice DF, Mackay CL, Goodwin RJ, McBride A, Langridge-Smith PR, Webster SP, Walker BR, Andrew R (2013) Mass spectrometry imaging for dissecting steroid intracrinology within target tissues. Anal Chem 85(23)11576–11584

53. Cooks RG, Ouyang Z, Takats Z, Wiseman JM (2006) Detection technologies. Ambient mass spectrometry. Science 311(5767)1566–1570

54. Wenk MR (2005) The emerging field of lipidomics. Nat Rev Drug Discov 4(7)594–610

Chapter 2

Nontargeted Lipidomics Utilizing Constant Infusion High-Resolution ESI-Mass Spectrometry

Paul Wood

Abstract

High-resolution (<3 ppm mass error) ESI-mass spectrometric analyses of lipid extracts allow for the analysis of over 2700 individual lipids across 56 lipid subfamilies utilizing a 1 min constant infusion. This approach allows broad sampling of the lipidome while minimizing issues of lipid losses, of racemization with silica-based supports, and of ghost effects that are associated with almost all chromatographic systems for lipids containing very-long-chain fatty acids. The issue of isobars must always be assessed carefully and can be managed by tandem mass spectrometry in some cases, by derivatization prior to direct infusion analyses, or via the use of chromatographic systems.

Key words High-resolution mass spectrometry, Lipid nomenclature, Tandem mass spectrometry

1 Introduction

Lipidomics platforms are incredibly valuable in that they allow investigators to interrogate precious clinical samples in addition to animal models of disease. This approach is utilized by investigators to examine biofluids and tissues from different disease populations and thereby determine which animal models are optimal for evaluating experimental treatment paradigms prior to clinical trials. This approach also is invaluable in establishing lipid biomarkers of disease, of disease progression (temporal axis), of the therapeutic response in patients, for defining patient subpopulations (patient stratification at the molecular level), and for defining new molecular therapeutic targets.

While lipidomics is a rapidly expanding sector of the "omics" technologies, it is essential to understand that the early work in this field was conducted with triple quadrapole mass spectrometers.

Electronic supplementary material: The online version of this chapter (doi:10.1007/978-1-4939-6946-3_2) contains supplementary material, which is available to authorized users.

Paul Wood (ed.), *Lipidomics*, Neuromethods, vol. 125,
DOI 10.1007/978-1-4939-6946-3_2, © Springer Science+Business Media LLC 2017

Therefore, data was obtained with unit mass resolution (50–200 ppm mass error) resulting in the potential for miss-assignments of lipid identities. The introduction of commercial high-resolution mass spectrometers has provided the technology to correct potential errors and to further advance the field of lipidomics. In our Lipidomics Unit, we utilize high-resolution ESI-mass spectrometry [1] for the analysis of diverse clinical samples and samples from transgenic mouse models. In the case of Mild Cognitive Impairment (MCI) and Alzheimer's disease, we have demonstrated abnormal accumulation of diacylcglycerols, early in the disease process, in brain tissue [1–3], CSF [1], and plasma [3, 4]. In addition, examining larger patient populations has allowed us to stratify these patient populations into at least three subgroups [5]. Similarly, we have demonstrated abnormal lipid metabolism in the myelin of brain samples obtained from subjects with schizophrenia and in a transgenic mouse model of this disease [6, 7]. Alterations of complex lipids in gray matter also were detected in the brain samples obtained from subjects with schizophrenia [8].

2   Materials and Equipment

All organic solvents should be of LC-MS grade.

2.1   Internal Standards

- Internal standards are made up as 1 mM solutions in methanol and are stored at −20 °C. Chloroform and/or water are utilized with lipids insoluble in methanol alone.

- All standard solutions should be warmed to room temperature before removing aliquots for the internal standard cocktail, since some of the lipids precipitate at −20 °C.

- Bromocriptine is included as an internal standard to monitor for potential mass axis drift and to evaluate isotopic resolution (Br^{81} = 49.3%). Bromocriptine can be monitored both in positive ESI and negative ESI.

- The following is a table of the internal standards we use to analyze 100 μL samples of human plasma (nmoles = nanomoles of internal standard per 100 μL plasma):

Int. Std.	Supplier	nmoles	Exact Mass	[M+H]+
PtdC 28:0 [D54]	Avanti	1	729.8257	730.8330
PtdC 32:0 [D62]	Avanti	0.5	793.9337	794.9410
PtdC 34:1 [D31]	Avanti	4	790.7720	791.7793
MAG 18:1 [D5]	CDN	1	361.3240	362.3313
Cer [D31]	Avanti	1	568.7070	569.7143

(continued)

Bromocriptine	Sigma	0.5	653.2213	654.2285
Bromocriptine-81			655.2192	656.2265
Int. Std.	*Supplier*	*nmoles*	*Exact Mass*	*[M+NH4]+*
DAG 36:2 [C3]	Larodan	1	623.5480	641.5824
TAG 48:0 [D5]	CDN	3	811.7677	829.8021
Int. Std.	*Supplier*	*nmoles*	*Exact Mass*	*[M−H]−*
PtdE 34:1 [D31]	Avanti	1.2	747.7191	746.7119
PtdC 28:0 [D54]	Avanti	0.5	687.7783	686.7710
PtdS 34:1 [D31]	Avanti	4	791.0837	790.7011
PG 32:0 [D62]	Avanti	0.25	784.8990	883.8917
PG 34:1 [D31]	Avanti	0.5	779.7200	778.7127
Stearic acid [C18]	CIL	0.15	302.3318	301.3246
Arachidonic acid [D7]	Cayman	0.1	312.2905	311.2832
DHA [D5]	Cayman	0.5	333.2716	332.2643
VLCFA 26:0 [D4]	CDN	1.25	400.4218	399.4146
DC 16:0 [D28]	CDN	1.5	314.3902	313.3829
Cholesterol sulfate [D7]	Avanti	1	473.3556	472.3483
PA 34:1 [D31]	Avanti	1	704.6763	703.6690
Bromocriptine	Sigma	0.5	653.2213	652.21398
Bromocriptine-81			655.2192	654.2119

2.2 Infusion Solvent

The infusion solvent is 80 mL 2 propanol + 40 mL methanol + 20 mL chloroform + 0.5 mL water containing 164 mg ammonium acetate.

2.3 Mass Spectrometer

The utility of the orbitrap mass spectrometer for direct infusion lipidomics studies was first reported by Schuhmann et al. [9]. The advantages of the Q Exactive (Thermo Fischer) for lipidomics analyses include:

1. A benchtop quadrupole orbitrap mass spectrometer with a small footprint.
2. High mass resolution (140,000).
3. High mass accuracy (0.4–3 ppm).
4. High sensitivity.
5. Excellent stability (no mass axis drift over 24 h. experimental period, with over 3 years of instrument operation).
6. Maintenance-free analyzer [8].

3 Methods

3.1 Extraction of Biofluids

1. 100 µL of biofluid is mixed with 1 mL of methanol containing the internal standard mixture and 1 mL of water [4].
2. Next 2 mL of t-butylmethyether is added and the tubes capped and shaken vigorously at room temperature for 30 min using a Fisher Multitube Vortexer.
3. The tubes are then centrifuged at $3000 \times g$ for 15 min at room temperature.
4. Next 1 mL of the upper organic lipid layer is transferred to a screw-top 1.5 mL microfuge tube and the samples dried for 3.5 h in an Eppendorf Vacufuge.
5. The samples are redissolved in 150 µL of the Infusion Solvent and the tubes centrifuged at $30,000 \times g$ for 5 min in an Eppendorf Microfuge to precipitate any particulates.
6. Note: This extraction procedure is not optimal for extraction of gangliosides.

3.2 Extraction of Tissues and Tissue Culture Cells

1. In the case of tissues [10] and cells, they are first sonicated in the methanol–water solution, as above, prior to the addition of the t-butylmethyether.

3.3 Mass Spectrometric Workflow Analysis

1. The redissolved sample is infused into the ESI source at a flow rate of 5 µL/min.
2. The sample is scanned first in Neg-ESI (180–1400 amu) mode for 0.5 min followed by Pos-ESI (200–1400 amu) mode for 0.5 min.
3. Both modes were set to scan at a resolution of 140,000.
4. The syringe and infusion line are washed between infusions first with 1 mL of methanol and then 1 mL of hexane:ethyl acetate (3:2) to minimize memory effects.

3.4 Mass Spectrometric Data Reduction

For the high-resolution mass spectrometric data, the top 1000 masses, using 200 amu windows, and their associated ion intensities are electronically transferred to an Excel spreadsheet. Within the spreadsheet is a list of 2689 lipids, their exact masses (five decimals), and the scanned ion which is searched within the data table. If the calculated ppm mass error is determined to be ≤ 3 for the extracted anion or cation, then the ratio of the ion intensity to that of the assigned internal standard is calculated and added to the spreadsheet. The following is a list of the 56 lipid subclasses that we monitor via direct infusion analyses.

3.5 Lipid Classes

The following is a list of the 56 lipid subclasses that we monitor via direct infusion analyses.

3.5.1 Glycerophos-
pholipids

The glycerophosphospholipid (GPL) subclasses and the associated ionization modes include:

1. Diacyls
    (a) Phosphatidylcholines (PtdC): [M+H]⁺
    (b) Phosphatidylethanolamines (PtdE): [M–H]⁻
    (c) Phosphatidylserines (PtdS): [M–H]⁻
    (d) N-Acyl-Phosphatidylserines (NAPS): [M–H]⁻
    (e) Phosphatidylinositols (PtdI): [M–H]⁻
    (f) Phosphatidylglycerols (PtdG): [M–H]⁻ and [M+H]⁺
    (g) Bis(monoacylglycero)phosphates (BMP): [M–H]⁻ and [M+H]⁺
2. Alkyl-Acyls (OA)
    (a) OA-glycerophosphocholines: [M+H]⁺
    (b) OA-glycerophosphoethanolamines: [M–H]⁻
3. Alkenyl-Acyls (Plasmalogens)
    (a) Choline plasmalogens (PlsC): [M+H]⁺
    (b) Ethanolamine plasmalogens (PlsE): [M–H]⁻
4. Lyso(Acyl)-GPL
    (a) Lysophosphatidylcholines (LPC): [M+H]+
    (b) Lysophosphatidylethanolamines (LPE): [M–H]–
    (c) Lysophosphatidylserines (LPS): [M–H]–
    (d) Lysophosphatidylinositols (LPI): [M–H]–
    (e) Lysophosphatidylglycerols (LPG): [M–H]–
    (f) Formyl-LPC: [M+H]+
    (g) Formyl-LPE: [M–H]⁻
5. Lyso(Alkyl)-GPL
    (a) Lyso-alkylphosphatidylcholines (LPCe): [M+H]+
    (b) Lyso-alkylphosphoethanolamines (LPEe): [M–H]⁻
6. Lyso(Alkenyl)-GPL
    (a) Lyso-alkenylphosphocholines (LPCp): [M+H]+
    (b) Lyso-alkenylphosphoethanolamines (LPEp): [M–H]⁻

3.5.2 Glycerols

The glycerol subclasses include:

1. Monoacylglycerols (MAG): [M+H]⁺
2. Diacylglycerols (DAG): [M+NH$_4$]⁺
3. Triacylglycerols (TAG): [M+NH$_4$]⁺

4. TAG hydroperoxides: $[M+NH_4]^+$

5. Alkylacylglycerols (EAG): $[M+NH_4]^+$

6. Alkenylacylglycerols (VAG): $[M+NH_4]^+$

7. Seminolipids (SEM): $[M-H]^-$

3.5.3 Sphingolipids

The sphingolipid subclasses include:

1. Sphingomyelins (SM): $[M+H]^+$

2. Galactosylceramides: $[M-H]^-$

3. Lactosylceramides: $[M-H]^-$

4. Ceramides (CER): $[M+HCOO]^-$

5. Ceramide-phosphoethanolamines (CER-PE): $[M-H]^-$

6. Phosphatidic acids (PA): $[M-H]^-$

7. Lysophosphatidic acids (LPA): $[M-H]^-$

8. Cyclic Phosphatidic acids (cPA): $[M-H]^-$

9. Sulfatides: $[M-H]^-$

10. Cardiolipins: $[M-2H]^{2-}$

3.5.4 Fatty Acids and Fatty Acid Derivatives

The fatty acid subclasses include:

1. Free fatty acids (FFA): $[M-H]^-$

2. Very-long-chain fatty acids (VLCFA): $[M-H]^-$

3. Hydroxy fatty acids (HFA): $[M-H]^-$

4. Dicarboxylic fatty acids (DC): $[M-H]^-$

5. Hydroxy dicarboxylic fatty acids (HDC): $[M-H]^-$

6. Dihydroxy dicarboxylic fatty acids (DHDC): $[M-H]^-$

7. Isoprenoids: $[M-H]^-$

8. Acyl-Carnitines: $[M+H]^+$

9. N-Acylserines: $[M-H]^-$

10. O-Acyl-hydroxy fatty acids (OAHFA): $[M-H]^-$

11. Cholesterol esters: $[M+NH_4]^+$

12. Cholesterol ester hydroperoxides: $[M+NH_4]^+$

13. Wax esters: $[M+NH_4]^+$

3.5.5 Sterol Sulfates

1. Cholesterol sulfate: $[M-H]^-$

2. Neurosteroid sulfates: $[M-H]^-$ and $[M-2H]^{2-}$

3. Bile acid sulfates: $[M-H]^-$ and $[M-2H]^{2-}$

3.5.6 Bacterial Fatty Acids

1. Mycolic acids: $[M-H]^-$

4 Tandem Mass Spectrometric Analyses

High-resolution mass spectrometric analyses provide a large amount of accurate lipidomics data. In many cases, tandem mass spectrometric analyses are also required to validate structures and to distinguish isobars [11]. A useful list of MS^2 spectra for different lipid classes generated with the Q-Exactive is included in Electronic Supplementary Material.

Acknowledgments

I wish to thank my basic and clinical research collaborators and the medical and masters students whose rotations in the Metabolomics Unit have aided in building and validating the lipidomics platform described in this chapter. In addition, this work would not have been possible without the financial support of Lincoln Memorial University, Rhizo Kids International, and the CFDA.

References

1. Wood PL (2014) Mass spectrometry strategies for targeted clinical metabolomics and lipidomics in psychiatry, neurology, and neuro oncology. Neuropsychopharmacology 39:24–33

2. Wood PL, Barnette BL, Kaye JA, Quinn JF, Woltjer RL (2015) Non-targeted lipidomics of CSF and frontal cortex grey and white matter in control, mild cognitive impairment, and Alzheimer's disease subjects. Acta Neuropsychiatr 27:270–278

3. Wood PL, Medicherla S, Sheikh N, Terry B, Phillipps A, Kaye JA, Quinn JF, Woltjer RL (2015) Targeted lipidomics of fontal cortex and plasma diacylglycerols in mild cognitive impairment (MCI). J Alzheimers Dis 48:537–546

4. Wood PL, Phillipps A, Woltjer RL, Kaye JA, Quinn JF (2014) Increased lysophosphatidyl-ethanolamine and diacylglycerol levels in Alzheimer's disease plasma. JSM Alzheimers Dis Relat Dement 1:1001

5. Wood PL, Locke VA, Herling P, Passaro A, Vigna GB, Volpato S, Valacchi G, Cervellati C, Zuliani G (2015) Targeted lipidomics distinguishes patient subgroups in mild cognitive impairment (MCI) and late onset Alzheimer's disease (LOAD). BBA Clin 5:25–28

6. Wood PL, Filiou MD, Otte DM, Zimmer A, Turck CW (2014) Lipidomics reveals dysfunctional glycosynapses in schizophrenia and the G72/G30 transgenic mouse. Schizophr Res 159:365–369

7. Wood PL, Holderman NR (2015) Dysfunctional glycosynapses in schizophrenia: disease and regional specificity. Schizophr Res 166:235–237

8. Wood PL (2014) Accumulation of N-acylphosphatidylserines and N-acylserines in the frontal cortex in schizophrenia. Neurotransmitter (Houst) 1:e263

9. Schuhmann K, Almeida R, Baumert M, Herzog R, Bornstein SR, Shevchenko A (2012) Shotgun lipidomics on a LTQ Orbitrap mass spectrometer by successive switching between acquisition polarity modes. J Mass Spectrom 47:96–104

10. Wood PL, Shirley NR (2013) Lipidomics analysis of postmortem interval: preliminary evaluation of human skeletal muscle. Metabolomics 3:3

11. Michalski A, Damoc E, Hauschild JP, Lange O, Wieghaus A, Makarov A, Nagaraj N, Cox J, Mann M, Horning S (2011) Mass spectrometry-based proteomics using Q Exactive, a high-performance benchtop quadrupole Orbitrap mass spectrometer. Mol Cell Proteomics 10:M111.011015

Ultrahigh-Resolution Lipid Analysis with Fourier Transform Ion Cyclotron Resonance Mass Spectrometry

Mark R. Emmett and Cheryl F. Lichti

Abstract

Global lipid analysis/profiling (lipidomics) is a rapidly expanding field that is a focal point of integrated omics (systems biology) studies of disease. Integrated omics is defined as the correlation of interdisciplinary technologies to interpret the molecular interactions within a biological organism, in response to gene/chemotherapy or disease state, by monitoring and correlating: phenomics, genomics, transcriptomics, lipidomics, proteomics, PTM profiling, glycomics, metabolomics, and bioinformatics.

Key words FT-ICR, Ultrahigh-resolution mass spectrometry, Gangliosides, Ionization modes

1 Introduction

Lipidomic analysis is ground zero in integrated omics studies due to the high abundance of and rapid turnover of lipids in the biological system. Since lipids are not coded for at the genetic level, but instead are enzymatically derived, lipids are tightly linked to the proteins (enzymes) that are crucial to their synthesis and degradation. Lipids are the "window" to the transcriptomic (coding) and proteomic (functionality) systems of the biological organism. These systems are tightly linked and can be used to assemble predictive molecular biochemical pathways involved in the diseased state as compared to normal control. This interactive data is efficiently used to intelligently design successive experiments, identify potential therapeutic targets and biomarkers, and rapidly enhance our knowledge of the molecular mechanisms involved in the initiation and proliferation of disease.

Lipidomics is currently grouped in the field of metabolomics, although lipids play a much greater role in the biological system than just as a metabolite. Lipids are biochemically important for many reasons: lipids are important structural molecules maintaining the integrity of the cell membrane, lipids are energy storage molecules, lipids are involved in cell adhesion, in cell/cell communication

Paul Wood (ed.), *Lipidomics*, Neuromethods, vol. 125,
DOI 10.1007/978-1-4939-6946-3_3, © Springer Science+Business Media LLC 2017

as ligands/receptors, and as second messengers linked to cell cycle, proliferation, survival, and signal transduction. Lipids have been linked to many diseased states: cancer, neurodegeneration (i.e., Alzheimer's, Parkinson's, Huntington's), and diabetes, to name a few [1, 2]. Glycosphingolipids are involved in several aspects of cancer proliferation, apoptosis, invasion, and metastasis [3].

The analysis of lipids is an analytical challenge due to their diverse structural nature. Most are amphipathic molecules with a polar head and nonpolar hydrocarbon tail, but others are highly nonpolar. The diversity of lipids skyrockets when one considers modifications such as degree of saturation, location of double bonds, and oxidations. It has been estimated that there are more than 180,000 possible combinations of phospholipids alone [4]! There is no "silver bullet" currently available for either extraction or analysis; their amphipathic nature makes both analysis and extraction equally challenging.

The focus of this book is on the analysis of biological lipids, the majority of which is performed by mass spectrometry (MS). That being said, there is still much work being performed utilizing thin-layer chromatography (TLC)–even coupling TLC with MS [5]. Nuclear magnetic resonance (NMR) is also used for lipid analysis [6], although NMR sensitivity is much less than that of MS. There are many types of mass analyzers and ionization sources available for biological lipid analysis of which several will be presented. As with extraction methods, there is no single MS method/instrument that can accomplish all analyses. Each method has limitations, and the researcher must realize the benefits and limitations of the instrumentation available for their research. The emphasis of this chapter will be on the applications of ultrahigh-resolution mass analysis of lipids by Fourier transform ion cyclotron resonance mass spectrometry (FT-ICR MS).

2   Lipid Extraction

Extraction of lipids from biological samples is really the second step in the analysis of lipids. The first step is correct handling of the biological sample prior to extraction to limit degradation of the lipids in the sample. Rapid freezing of the sample should occur as soon as possible after harvest. This is easily done with samples harvested from animals or tissue culture, but sample handling is a factor that the researcher must take into consideration when dealing with postmortem samples. Assuming that the samples have been collected in a manner to limit degradation, tissues and bio-fluids (blood, urine, saliva, sputum, etc.) are very complex mixtures containing proteins, cells, metabolites, salts, carbohydrates, mucins, etc. and contain a multitude of lipid molecules with a wide range of polarity and concentrations [7]. Researchers working with brain samples or tissue culture derived from brain [8, 9] have the easiest task because the brain is the most lipid abundant organ in the body [2]. Brain may have the highest concentrations of lipids, but lipids have been extracted and analyzed efficiently from a multitude of other sample types as reflected in several recent publications: cancerous tissue [10], tissue culture [3], retina [1], cerebrospinal fluid (CSF) [9, 11], blood [12], exosomes [13], fish [14, 15], plants [16], algae [17, 18], and eukaryotic flagella [19]. This is by no means a complete list, but it demonstrates that lipids can be extracted and analyzed from most of all biological sources, and efforts are underway to analyze lipids on the single cell level [20].

There are basically two disciplines for the extraction of lipids from biological samples: solvent/solvent extraction and solid phase extraction (SPE). Within each of these disciplines, there are a multitude of variations. There is no one extraction method that guarantees extraction of all lipids in a biological sample due to the vast diversity of chemical composition, polarities, and size of lipids. Review of extraction methodologies is beyond the scope of this chapter, thus a short summary will follow. For more detailed information, review Reis et al. [21], Wasntha et al. [22], and Lydic et al. [1] for solvent/solvent extractions and Bojko et al. [7] for SFE extraction. Raterink et al. [23] provide a short comparison of both extraction methods.

For solvent/solvent extractions there are three predominant methods: Bligh/Dyer [24], Folch [25], and MTBE [26], and multiple variations thereof. The first two utilize chloroform and methanol as their organic extraction solvents and achieve a phase separation with the addition of water. The third uses methyl tert-butyl ether (MTBE) and methanol as the organic solvents with the addition of water to induce a phase change. Each of these protocols has its pros/cons, yet none seems to equally extract all lipids. The MTBE extraction has advantages of being less toxic than the first two and also allows the recovery of an intact protein pellet, which can be used to equate lipid levels back to total protein in the sample. In our lab, we predominately use a modification of the Bligh/Dyer method because it yields more of the larger lipids such as gangliosides, which are a focus of our research, and it works well with small sample size (we routinely extract from 20 μm thick mouse whole brain slices). The MTBE extraction is convenient, but in our hands it does not extract the larger gangliosides as well as the Bligh/Dyer. In all solvent/solvent extractions it is a good idea to add 1 mM butylatedhydroxytoluene (BHT) to the extraction solvents to reduce oxidation of lipids during the extraction process.

SPE is a very attractive extraction method for lipids due to its ease of use and adaptation to automated high-throughput extraction and analysis. The main problem for global lipidomics is that SPE is highly selective for lipid classes and suffers from poor recovery [7, 27, 28]. With that being said, SPE is the method of choice for targeted lipid analysis.

Finally, two other intriguing extraction methodologies that may hold great benefit are electromigration/electroelution [23] and supercritical fluid extraction (SFE) [29]. Electromigration extraction shows great promise because it pre-concentrates the analyte during the extraction procedure and is also amendable to automation. The majority of the publications to date have focused on extraction of small metabolites; the effectiveness of this method for lipid extraction is yet to be demonstrated. Due to the varying charges on lipid molecules, it makes logical sense that this extraction method would be most suited to targeted analysis. SFE holds great promise for extraction of lipids from biological samples. In SFE, supercritical gas (most often supercritical CO_2) is used as the extraction solvent. Supercritical CO_2 is considered to be a nonpolar solvent, but it also has polar characteristics due to its large molecular quadrupole [30]. The polarity of the supercritical CO_2 can easily be increased with the addition of modifiers such as methanol or other polar solvents, and thus both polar and nonpolar compounds can be extracted simultaneously. SFE is efficient and fast but have been hampered by high volume extraction vessels (1–24 mL and higher). Clever researchers have miniaturized the extraction vessel by performing SFE in an empty HPLC guard column [29]. A commercial vendor (Shimadzu Scientific Instruments, Inc.) has introduced an SFE vessel with a volume of 250 μL, which should be adequate for most biological extractions. The literature is sparse on the subject of SFE of lipids from biologicals; this is an area of research that holds great promise in lipid analysis and needs to be further explored.

3   Mass Spectrometry Analysis

After an extraction of global lipids, the sample contains a complex mixture of compounds of diverse polarities, molecular weights (many that are isobaric or isomeric), and modifications. As an example, consider the ganglioside family, which consists of lipids with highly varied polar sugar head groups coupled to a nonpolar ceramide tail that is also highly variable (length of ceramide tail, degree of and location of saturation, and extent of modification, i.e., oxidations). The complexity of gangliosides makes them a challenge to analyze by MS. The sugar head group is often capped with a negatively charged sialic acid, which affects ionization. Additionally, the multiple branching of the sugar head groups results in many structural isoforms that often cannot be distinguished by MS or MS/MS methodology. Due to the complexity of the sample, there are many options in the MS analysis of lipids. Raterink [23] summed it up in his metabolomics review, "no single analytical technique covers the entire spectrum of the human metabolome; various complementary analytical platforms (e.g., gas chromatography-MS [GC/MS], liquid chromatography-MS [LC/MS], capillary electrophoresis-MS [CE/MS] and/or direct infusion-MS) should be employed in order to improve metabolite coverage and identification power." This same mentality applies to complex lipid analysis and will be expanded to include: ionization type, direct infusion or separation prior to ionization, MS instrumentation type (low or high resolution), MS/MS fragmentation types, and high-resolution data analysis. Each of these areas will be discussed in this section.

3.1   Ionization Modes   There are a wide variety of ionization modes available in MS. Some are mostly specific to the particular MS instrumentation, e.g., electron impact ionization (EI) in GC/MS and matrix-assisted laser desorption (MALDI) in many time of flight (TOF) MS instruments. Each ionization mode has advantages and drawbacks, and there is no single ionization mode that accomplishes all goals. The researcher is limited to what is available in his/her laboratory and must utilize that ionization mode to its fullest potential.

EI is the major ionization technique used when coupling a GC to a MS. EI is an efficient technique for ionization of molecules in the gas phase and is therefore highly suitable for molecules eluting from a GC column. In a nutshell, the eluted molecules pass through a high energy electron beam (~70 eV) that strips an electron from the molecule, resulting in a positively charged radical ion. EI also results in significant fragmentation of the analyte, which can be beneficial for known compounds that are in a library, but can make interpretation of EI spectra of complex unknowns extremely difficult. For GC separation, analytes need to be volatile to make it through the GC column, so most compounds need to be derivatized prior to injection. The long carbon tails and increased molecular weight of many lipids limit their ability to be eluted from GC columns; therefore, GC/MS is primarily limited to smaller lipid compounds.

MALDI ionization [31, 32] was introduced as a "soft" ionization technique. MALDI utilizes a "matrix" that is mixed with or applied to the sample. The sample is hit with a laser beam, and the matrix absorbs the majority of the energy of the laser and ionizes. The laser beam desorbs both the analyte and ionized matrix into the gas phase, where the ionized matrix transfers a charge to the analyte, forming an ion for MS analysis. MALDI can be interfaced to a variety of mass analyzers of varying resolutions. MALDI has the benefit of being highly tolerant of salts in the sample and has been used extensively for MALDI imaging [33] (primarily of lipids), which provides localization of specific lipid ions produced from a tissue slice. In this technique, the MALDI matrix is applied to the tissue, and individual laser shots are made incrementally across the tissue sample. A mass spectrum is collected at each spot, and the multiple spectra are assembled to reveal a molecular ion image of the tissue. It is important to note that interpretation of MALDI and MALDI imaging results for biological lipids should be performed carefully due to the loss of labile sialic acid residues during the MALDI ionization process [34]. It is anticipated that smaller beam size lasers (not yet available) will allow imaging at the single cell level. Recent efforts by Muddiman's group [20], using an oversampling technique with a hybrid MALDI-electrospray source (IR-MALDIESI), generated data that predicts the identification of cholesterol at the single cellular level.

Electrospray ionization (ESI) was introduced by John Fenn's group in 1985 [35]. Since this ionization mode is operated at atmospheric pressure, it allowed the convenient interfacing of LC to MS. Even more importantly, ESI produced multiply charged ions, which had the effect of greatly extending the mass range of all MS systems to which it was interfaced. Since MS systems monitor the mass-to-charge ratio (m/z) of ions, rather than simply monitoring mass, ESI truly allowed the analysis of large biomolecules for the first time. Due to the significance of this advancement, John Fenn shared the Nobel Prize in Chemistry in 2002 for the invention of ESI. The advent of low-flow ESI techniques (micro-ESI [36] and nano-ESI [37]) revolutionized biological mass spectrometry due to the increase in sensitivity by multiple orders of magnitude as compared to conventional ESI. The increased sensitivity is due to increased ionization efficiency, which is the result of enhanced desolvation of the smaller charged droplets produced by low-flow ESI and elimination of sheath flow and/or sheath gas. The only real difference between micro-ESI and nano-ESI is that, with micro-ESI, the flow through the capillary emitter is pumped (syringe pump, HPLC pump, air pressure, etc.); whereas with true nano-ESI, flow rate is dependent on the potential applied to the emitter tip [38, 39]. Most often in micro-ESI, voltage is applied to the solvent, and the spray is generated from a nonconductive capillary emitter tip. The nonconductive tip eliminates arcing in the negative ion mode, thus eliminating the need for scavenging gases such as O_2 or SF_6 and allows for easy switching between positive and negative ion modes. The majority of lipid analysis with MS is performed with low-flow ESI sources.

Other atmospheric sources that have major applications in lipid analysis (and can readily interface LC to the MS), but tend to be underutilized are: atmospheric pressure chemical ionization (APCI) and atmospheric pressure photoionization (APPI). APCI is based on an original design by Horning [40] in which the solvent is infused at high flow through a heated inlet and passed through a corona discharge for ionization (in the presence or absence of reagent gas). Henion later utilized APCI for efficient

LC-MS/MS applications [41]. A major advantage of APCI for lipid analysis is the increased ionization efficiencies of nonpolar compounds. APPI [42] is especially applicable for ionization of nonpolar compounds and is underutilized in nonpolar lipid analysis. In APPI, a high energy photon is absorbed by either the analyte or the solvent, which causes the ejection of an electron, resulting in the production of a radical cation ($M\bullet^+$). This radical cation can then donate a proton to the analyte, producing a stable cation $[M+H]^+$. Negative ions can also be produced by creating an abundance of thermal electrons. An advantage of APPI is that it does not suffer from ion suppression to the same extent as ESI or APCI [43]. It is important to note that APPI produces both stable and radical cations, which complicates mass spectral analysis. For a review on the comparison of ESI, APCI, and APPI for lipid analysis, consult Imbert et al. 2012 [44]. In summary, ESI is optimal for polar lipid analysis, but has diminished response on nonpolars. APCI has better response on nonpolar lipids than ESI and APPI has the best response on nonpolar lipids, but fails on polar lipid analysis.

Finally, there are ionization sources that combine lipid extraction with ionization and interfacing to the MS. Costello's group reported on two commercially available systems for extracting gangliosides from TLC plates, the CAMAG TLC interface (CAMAG Scientific, Inc.) and the liquid extraction surface analysis (LESA, Advion BioSciences) [5]. Both systems performed well for extraction and analysis of gangliosides from TLC plates. Desorption electrospray ionization (DESI) [45] has recently been demonstrated to lift lipids directly from human esophageal adenocarcinoma tissue samples for MS analysis [10].

In summary, there are many ionization modes available for biological lipid analysis. There is no one ionization method that has a stable response across all lipid polarities, so the researcher must compromise based on what source is available. If multiple sources are available and the extraction is a dual phase extraction such as the Bligh/Dyer, it would be advisable to use ESI analysis for the aqueous phase extraction and APCI and APPI analysis for the chloroform phase. Currently, ESI is the most common and versatile ionization source used in lipid analysis. Thus, the remainder of the discussions in this chapter will focus mainly on ESI applications. The next question is, should the analysis be performed in positive (pos.) or negative (neg.) ion mode? For global lipid analysis, neg. ion mode has been demonstrated to have superior sensitivity in lipid analysis [8, 27, 46]. However, for targeted lipid analysis, pos. ion analysis may provide optimal results for the lipid class of interest. If time and MS duty cycle permits, both pos. and neg. ion analysis would be preferred.

3.2 Sample Separations Prior to MS Analysis

There are two major methods for performing MS analysis on extracted lipid samples: direct infusion (shotgun) and sample separation (capillary electrophoresis [CE], LC, or supercritical fluid chromatography [SFC]) prior to MS analysis. Shotgun analysis is attractive due to the ease and speed of analysis and its generation of much smaller data sets. Addition of a chromatographic step prior to MS analysis requires additional equipment/expertise, adds time and complication to the analysis, and produces large data sets. There are pro/cons to each method and the method of choice boils down to what the researcher is trying to achieve in their analysis.

Direct infusion of lipid extracts is appealing due to the simplicity of the analysis, making shotgun lipidomics the method of choice in many research labs [9, 47–50]. The solvent/solvent biological lipid extractions produce extremely complex mixtures, and direct infusion of these mixtures necessitates analysis with high-resolution MS to help distinguish overlapping ions. Even with the highest resolution MS available, there are factors that must be considered when dealing with such a complex mixture. One often overlooked factor in the ESI-MS analysis of complex biological mixtures is that, as complexity of the sample increases, there is a competition for charge to be transferred to the analytes that are dissolved in the charged droplet produced by the low-flow ESI source. The charges on the droplet are finite and will be transferred to the ions that will most readily accept that charge. Thus, in a very complex biological sample, some analytes will rarely (if ever) pick up a charge and become ions. This ion suppression is a well-known complication in the ESI analysis of complex mixtures [51]. Reduction of the complexity of the sample is critical to achieve ionization of the maximum number of components in the complex biological mixture. This is the major reason that some form of separation prior to ionization is necessary in order to observe the most complete lipid profile of complex biological samples.

Figure 1 illustrates the importance of sample separation (in this case liquid chromatography) in the analysis of a complex biological lipid sample [8]. The goal of these experiments was the development of a method to extract and relatively quantify gangliosides from glioma tissue culture. Lipids were extracted by a modified Bligh/Dyer, and the extracted biological sample (aqueous phase) was spiked with four ganglioside standards: GM1 (d18:1/16:0), GM2 (d18:1/16:0), GM2 (d18:1/18:0), and GM3 (d18:1/16:0). The sample was directly infused and ionized by micro-ESI into a 14.5 T FT-ICR MS. The lower spectrum shows the response for the spiked ganglioside standards in the biological background with constant infusion. In the broadband spectrum, GM2 and GM3 are the only standards seen at low levels in the biological lipid background. The lower right inset spectrum shows a zoomed spectrum in the m/z mass range of 1425–1650; no GM1 standard is seen

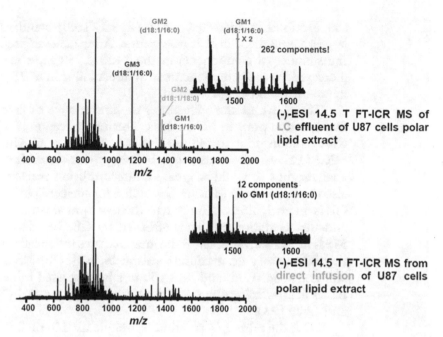

Fig. 1 Comparison of Constant Infusion vs. LC/MS of Biological Lipids. *Lower Spectrum*: Constant infusion of the aqueous phase of a Bligh/Dyer extraction of glioma cell culture after drying down and resuspension in 80% MeOH, 10 mM Ammonium Acetate. Sample spiked with GM1 (d18:1/16:0), GM2 (d18:1/16:0), GM2 (d18:1/18:0), and GM3 (d18:1/16:0) standards and directly infused by negative ion micro-ESI into a 14.5 T FT-ICR MS. Only GM2 and GM3 standards are detected in the biological lipid background. *Lower Right Inset Spectrum*: Zoomed spectrum m/z range of 1425–1650, no GM1 standard is seen above the biological background and only 12 endogenous components (features) were detected in the 1425–1650 m/z mass range with direct infusion. *Upper Spectrum*: Response of the above ganglioside spiked biological extract after nano-LC fractionation over a 50 μm i.d. × 8 cm nano-LC column in-house packed with phenyl-hexyl resin (Phenomenex, Torrance, CA). The full spectrum represents the sum of all chromatographic spectra to produce one spectrum for presentation purposes. Note the increased response of the ganglioside standards (including GM1) above the biological background after nano-LC separation. *Upper Right Inset Spectrum*: Zoomed spectrum m/z range of 1425–1650, after nano-LC separation, GM1 (d18:1/16:0) standard is the largest component in the m/z range and endogenous components increased from 12 components with constant infusion to 262 components after nano-LC separation

above the biological background. Also note that only 12 endogenous components (features) were detected in the 1425–1650 m/z mass range when the extract was constantly infused. The upper spectrum in Fig. 1 shows the signal response for the same ganglioside-spiked biological extract sample after nano-LC fractionation over a 50 μm i.d. ×8 cm nano-LC column, packed in-house with phenyl-hexyl resin (Phenomenex, Torrance, CA). To assemble the lower full spectrum, all of the chromatographic spectra were summed to produce one spectrum for presentation purposes. Even with the summing, it is easy to see the benefit of chromatographic separation because all of the spiked ganglioside standards are now readily apparent above the biological baseline. The upper inset of the zoomed 1425–1650 m/z mass range shows the previously undetected GM1 (d18:1/16:0) standard as the largest component in that m/z range. After nano-LC separation, the number of components in the 1425–1650 m/z mass range increased from 12 components with direct infusion to 262 components with nano-LC separation.

The above example illustrates the importance of lipid sample fractionation prior to MS analysis. The next question is what type of separation prior to MS analysis is best: SPE, GC, CE, LC, or SFC? SPE was discussed above as being more suited for targeted analysis; thus, it would be great for direct infusion analysis. GC was also discussed above as being best suited for smaller lipid molecules. CE is an intriguing option due to the low sample volumes needed and the separation efficiency afforded by CE. Interfacing CE to MS is not trivial, especially if the interface uses the sheathless design that is necessary for maximum sensitivity [52]. CE/MS analysis is still not mainstream and thus will not be discussed in depth; the basics of interfacing CE to MS can be found in a review by Maxwell and Chen [53].

LC is the primary separation technique used in the MS analysis of biological lipids, and there are many good references on the subject. Cajka and Fiehn recently published an extensive review on LC/MS of lipids [27]. The first consideration is what format LC to interface to MS: conventional, microbore or nano-LC. The LC configuration and flow rate will be dictated by the ion source available on the researcher's MS instrument. The highest sensitivity comes from nano-LC, but headaches such as column plugging, dead volumes, poor connections, and limited options of available resins (unless columns are packed in-house) accompany improved sensitivity.

The resin selection options for lipid analysis are vast. Biological lipid mixtures are a chromatographic challenge because the components range from very nonpolar to amphipathic. The most common resin used in lipid analysis is a reversed-phase (C-18 or C-8) resin [27], and this is reflected in the majority of the literature. Other resins employed in the literature are normal phase resins such as silica and hydrophilic interaction liquid chromatography (HILIC). Like all other aspects discussed previously, the selection of LC resin depends on speed of separation, family of lipid to be analyzed, size of lipid, etc. There is a vast array of resins available, but few researchers try resins other than those in the mainstream. C-18 resin is suitable for smaller lipids such as phospholipids, but will tightly bind lipids with long carbon tails. For example, gangliosides will often bind irreversibly to the resin, making C-18 unacceptable for ganglioside analysis. A good alternative is phenyl-hexyl resin (used in Fig. 1), which interacts primarily through π–π interactions. Phenyl-hexyl has good retention and release of a wide variety of biological lipids ranging from small sulfatides, phospholipids, and gangliosides up to C48. We have experimented with a new pentafluorophenyl (F5) resin (Phenomenex) that should have increased π–π interactions, resulting in better retention. As expected, retention was better–so much so that recovery was reduced. F5 resin has potential, but elution conditions need to be optimized. Schrader's group recently reported a silver ion (Ag^+)/ silica resin that also has strong π interactions [54]. This resin was used for the separation of asphaltenes, but it may also find use in the separation of biological lipids.

The lipid chromatography described in Fig. 1 was performed on the aqueous phase from the Bligh/Dyer extraction. Efficient chromatography of the nonpolar phase is a major challenge. The optimal separation method for these nonpolar lipids would be SFC, which is easily interfaced to MS. The supercritical CO_2 mobile phase will keep the nonpolar lipids in the solution, yet supercritical CO_2 also has polar characteristics that can be further increased by the addition of modifiers such as methanol. Thus, SFC has the capacity for the greatest range of lipids of any chromatographic technique. The capacity of SFC to handle polar compounds has been demonstrated by its use in the separation of highly polar peptide fragments in hydrogen/deuterium exchange experiments [55]. The Bamba [56–58] and Taylor groups [59] lead the field in SFC separations of lipids. Refer to Dr. Bamba's chapter in this book for more information on SFC/MS of lipids.

3.3　MS Instrumentation

All MS instrumentation performs the same basic function, which is to determine the mass of chemical analytes. Instruments vary in mass accuracy and resolution, ionization modes, sensitivity, ability to perform MS^n, detection modes, and duty cycle. There is no one MS instrument that outperforms all others; they each have their pros and cons. If you ask a researcher, "What is the best MS instrument?", you will get a wide range of answers, and most often their instrument of choice will be the instrument that is in their lab (which is the correct answer). The researcher is limited to the instrumentation that he/she has available to them, and it is up to them to use that particular instrumentation to its fullest potential in their research.

Mass accuracy and resolution are MS litmus tests. Mass accuracy is most often defined in parts per million (ppm) or parts per billion (ppb). A MS instrument that is experimentally accurate out to four decimal places on a molecular ion that has a m/z of 1000 has a mass accuracy of 1 ppb. Resolution, on the other hand, is the ability to resolve two closely placed peaks. Resolution is defined as the full width of a spectral peak at half-maximum peak height ($RP = m/\Delta m$). Example: a peak at m/z of 500 (m) and has a peak width at half height of 0.1 (Δm), then $500/0.1 = 5000$ resolution. Instruments that have high mass resolution also have higher mass accuracy if the MS is calibrated correctly. Instrument calibration is crucial; just because an instrument has high resolution does not necessarily mean that it has high mass accuracy. A typical quadrupole or ion trap has resolution of ~1000 and unit mass accuracy (1 Da), while time of flight instruments range from ~10,000 to 50,000 resolution. In this chapter, high-resolution instruments will be defined as those with resolution >100,000, which include Orbitrap and ultrahigh-resolution FT-ICR MS instruments. What the quadrupole, ion trap, and TOF instruments lack in mass accuracy and resolution, they make up in scan speed, which is much faster than the FT-MS instruments.

Both Orbitrap and FT-ICR MS instruments are Fourier transform (FT) instruments and detect ions based on their orbital frequencies in a trap. These detection modes are nondestructive, unlike electron multipliers in other mass spectrometers, which amplify the signal after the ion hits the detector. The nondestructive detection in the FT-ICR MS has unique benefits that will be discussed below. High-resolution lipid analysis with an Orbitrap will be addressed by Dr. Wood in another chapter in this book. The following is a discussion of the application of ultrahigh-resolution FT-ICR MS to lipid analysis.

3.3.1 Fourier Transform Ion Cyclotron Resonance Mass Spectrometry, FT-ICR MS

FT-ICR MS was coinvented by Alan G. Marshall and Mel B. Comisarow in 1974 [60]. FT-ICR MS instruments are the highest mass accuracy and highest resolution mass spectrometers available. These MS instruments utilize a high homogeneity magnetic field generated by a superconducting magnet. The ICR cell (an ion trap) is centered in the homogeneous magnetic field inside the super conducting magnet. In layman's terms, ions are generated outside the magnetic field, transferred into the field by ion guides (usually multipoles), and trapped in the ICR cell, in the high magnetic field and in high vacuum. Once trapped in the homogenous magnetic field, the ions will initiate their own cyclotron orbit within the stable magnetic field. The cyclotron frequency is related to the m/z of the ion. The ions are excited by a resonant excitation pulse, which pushes them closer to the detection plates of the ICR cell. Ions induce a charge on the detection plates of the ICR cell which is detected and collected as image current. The resulting data is then Fourier transformed to a frequency spectrum, which is mathematically converted to a m/z spectrum. This unique frequency-based detection system enables the unparalleled high mass accuracy and high resolution of FT-ICR MS. Mass accuracy is typically in parts per billion (ppb), and resolution can be several million depending on scan speed, FT-ICR cell design, and magnetic field. The most comprehensive review on FT-ICR MS theory, functions, and applications is authored by Marshall, Hendrickson and Jackson [61] and is commonly referred to as "The Primer." It is important to note that the performance parameters of FT-ICR MS scale with increased magnetic field. Some parameters increase linearly with increasing magnetic field (resolving power, scan speed) and others increase quadratically with increasing magnetic field (upper mass limit, ion trapping time, number of ions trapped). Thus, as magnetic field increases so does the performance of the FT-ICR MS. Commercially available systems range in magnetic field strengths from 7 to 15 Tesla (T). The highest performance instrument to date is the 21T FT-ICR MS at the National High Magnetic Field Lab (NHMFL) in Tallahassee, FL [62].

The strength of high resolution, high mass accuracy MS is in its ability to distinguish overlapping peaks in a complex mixture (e.g., monoisotopic peak or adducted peak of one compound overlapping with the ^{13}C isotope of another compound). Complex lipid mixtures contain many isobaric species (same nominal mass, but different chemical compositions) and there are several examples of the identification of isobaric lipid species with an Orbitrap [14, 50]. As an example, Schwudke et al. demonstrate the resolution of PS 36:2 and PC 36:1, which differs in mass by 726 mDa, on a high-resolution Orbitrap. This mass difference is easily resolved by FT-ICR MS. It should be noted that spectra can be greatly complicated by modifications and adducts often seen in biological phospholipids. For example, the monoisotope of PG (40:10) and the monoisotope of doubly oxidized PI (30:0) differ by only 5.9 mDa. Adduction of phospholipids by sodium or acetate can result in mass differences below 2 mDa between monoisotopic and ^{13}C isotopes of some phospholipid species.

The value of ultrahigh-resolution FT-ICR MS is readily apparent in the analysis of glycosphingolipids, which have many small mass isobaric species [3]. The smallest mass difference between two non-adducted monoisotopic glycosphingolipids (GSL) is 7.4 mDa, e.g., GD2 (d18:1/14:0) vs. SGPG (d18:1/26:2). Adducted GSLs (usually with sodium and acetate) greatly complicate the spectrum, as in the sodium adduct of GM3 (d18:1/22:0) and non-adducted GM3 (d18:1/24:3), which only differ by 2.4 mDa. The mass difference of the ^{13}C isotopes vs. the monoisotope of another GSL is even tighter, 1.6 mDa for the second ^{13}C isotope of 3-sulfoglucuronylparagloboside (d18:1/26:3) and the monoisotope of GD2 (d18:1/14:0)]. Only high-field, ultrahigh-resolution FT-ICR MS has the ability to resolve and identify phospholipid and glycosphingolipid compositions based solely on accurate mass measurement.

Another advantage of ultrahigh-resolution FT-ICR MS to aid in the identification of compounds is its ability to resolve isotopic fine structure. The monoisotopic peak of a compound is composed of the most abundant natural isotope for each atom that makes up its chemical composition (^{1}H, ^{12}C, ^{14}N, ^{16}O, ^{32}S, etc.), but each of these atoms also has at least one naturally abundant stable isotope (^{2}H [0.015%], ^{13}C [1.1%], ^{15}N [0.36%], ^{17}O [0.04%], ^{18}O [0.2%], ^{33}S [0.76%], ^{34}S [4.22%]). Due to the 4.22% natural abundance of the ^{34}S isotope, ultrahigh-resolution FT-ICR MS can easily split the ^{13}C$_2$ isotope and the ^{34}S isotope (difference of 10.9 mDa) in an

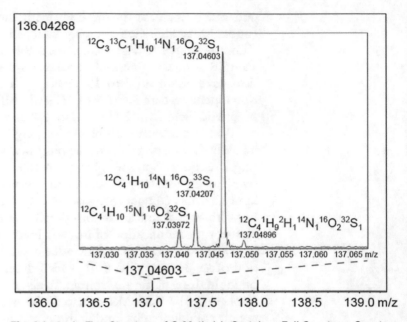

Fig. 2 Isotopic Fine Structure of S-Methyl-L-Cysteine. *Full Spectrum*: Spectrum of S-methyl-L-cysteine (C₄H₉NO₂S). Data collected at a LC compatible scan speed on a Bruker Solarix 12T FT-ICR MS equipped with an ultrahigh-resolution dynamically harmonized ParaCell™ ICR cell. Exact mass of [M+H]⁺ is 136.04267, experimental mass accuracy of 73 ppb. Spectrum shows [M−H]⁻ monoisotopic peak, the $^{13}C_1$ and $^{13}C_2$ isotope peaks. *Zoom Inset*: Zoomed spectrum of the first $^{13}C_1$ isotope on a 35 mDa m/z range. The $^{13}C_1$ isotopic peak is actually composed of four baseline resolved peaks (illustrating isotope fine structure). The natural isotope abundances are: ^{15}N-0.36%, ^{33}S-0.76%, ^{13}C-1.1%, and ^2H-0.015%. Exact mass, isotopic fine structure spacing, and abundances can be used to unambiguously assign elemental composition

ion that contains a sulfur atom. This verifies the presence of sulfur in the molecule without performing MS/MS. Shi et al. used the $^{13}C_2$ isotope and the ^{34}S isotope split to determine the number of sulfur atoms in a 16 kDa protein [63]. The potential value of isotopic fine structure afforded by ultrahigh-resolution FT-ICR MS is demonstrated in Fig. 2, which shows the spectrum of S-methyl-L-cysteine. The inset of Fig. 2 shows a zoomed spectrum of the first $^{13}C_1$ isotope on a 35 mDa m/z range. Note that the $^{13}C_1$ isotopic peak is actually composed of four baseline resolved peaks. The first peak (from left to right) corresponds to the chemical composition of S-methyl-L-cysteine, but contains one ^{15}N, the second peak contains one ^{33}S, the third peak is the actual ^{13}C containing peak, and the fourth peak contains one ^2H. This data was collected with an LC compatible scan speed on a 12T Bruker Solarix FT-ICR MS equipped with an ultrahigh-resolution dynamically harmonized ParaCell™ ICR cell. Just utilizing the sub-ppm mass accuracy of ultrahigh-resolution FT-ICR MS, elemental composition can most often be identified based solely on mass, but due to the large number of possible atomic configurations for a given nominal mass this is not always the case. By using the heteroatom content of the ion, the nominal spacing of the isotopes, the abundances of the isotopic peak, along with the ppb accurate mass, elemental composition can be assigned unambiguously without MS/MS [64].

Even with ppb mass accuracy and isotopic fine structure capabilities of ultrahigh-resolution FT-ICR MS, identification of true structural isomers (two compounds with the exact same elemental composition) requires MS/MS to aid in the determination of structural assignment. FT-ICR MS excels in MS/MS capabilities, some of which are unique to FT-ICR MS instrumentation. The nondestructive detection of ions in the FT-ICR trap allows ions to be detected in the cell, then fragmented in the FT-ICR cell (by a variety of fragmentation techniques), and the fragment ions are detected at ultrahigh-resolution. A fragment ion can be isolated, fragmented again and the fragment ions detected again. The MS^n capabilities of FT-ICR MS are limited by ion number and time of the MS^n experiments. This is a unique capability of FT-ICR MS in the realm of high-resolution instruments. Orbitraps have no capability to fragment ions once they are trapped in the Orbitrap. Most FT-ICR MS instruments have several options for MS/MS both prior to introducing the ions to the FT-ICR cell (external MS/MS) and after the ions are trapped (internal MS/MS). External MS/MS options are: post-source decay, collision-induced dissociation (CID), and electron transfer dissociation (ETD). Internal MS/MS options are: sustained off resonance irradiation (SORI), infrared multiphoton dissociation (IRMPD), electron capture dissociation (ECD), electron detachment dissociation (EDD), and electron-induced dissociation (EID). Combinations of these MS/MS techniques (external and internal) are routinely used in biological lipid analysis because the different techniques often produce unique fragmentation patterns. McFarland et al. demonstrated the different fragmentation of the ganglioside GM1 by comparing IRMPD and ECD on the positive ion of GM1 and IRMPD and EDD on the negative ion of GM1 [65].

The importance of multiple MS/MS fragmentation techniques (one unique to FT-ICR MS) in the analysis of biological lipids is demonstrated in Figs. 3–5 [66]. In the global lipid analysis of a glioma cell line, a highly abundant singly deprotonated negative ion of what was thought to be asialo-GM1 was preliminarily identified with low ppb mass accuracy. External CID in a low-resolution ion trap (hybrid instrument that coupled an ion trap to a 14.5 T FT-ICR MS) revealed that the ion was in fact either iGg4 or Gb4. Figure 3 shows the structure of the polar sugar head groups of asialo-GM1 and iGb4/Gb4 gangliosides. The sugar head group contains the same sugar residues in both, but in a different sequence in asialo-GM1 vs. iGb4/Gb4. This posed a new problem, because

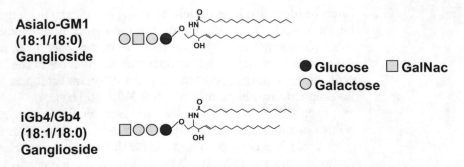

Fig. 3 Structures of Asialo-GM1 and iGb4/Gb4 Gangliosides. Both gangliosides contain the same composition of sugars in their head groups, but each have a different carbohydrate sequence in their head groups, thus their monoisotopic masses are identical

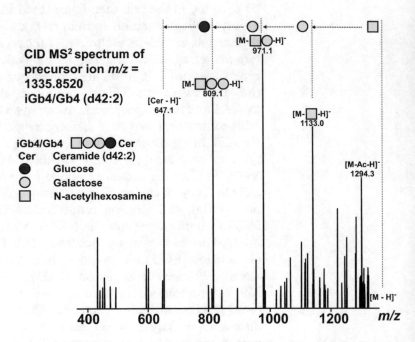

Fig. 4 MS/MS of [M−H]⁻ m/z 1335.852 Ion. MS/MS verifies that the biological lipid negative ion at m/z 1335.852 is either iGb4 or Gb4 and not asialo-GM1

the difference between iGb4 vs. Gb4 is the sugar linkage between the two galactose residues in the center of the polar sugar head group: iGb4 has a 1–3 linkage and Gb4 has a 1–4 linkage. In order to determine the sugar linkage, a MS/MS fragmentation technique was needed that efficiently caused cross-ring cleavages. This was complicated because the ion was a singly deprotonated negative ion, thus EDD was not an option (EDD necessitates multiply charged ions). To accomplish this, a technique developed in the Amster group, electron-induced dissociation (EID) [67] for dissociation of singly deprotonated ions, was utilized. EID conditions were first optimized on a Gb4 (d42:0) standard; the EID MS/MS

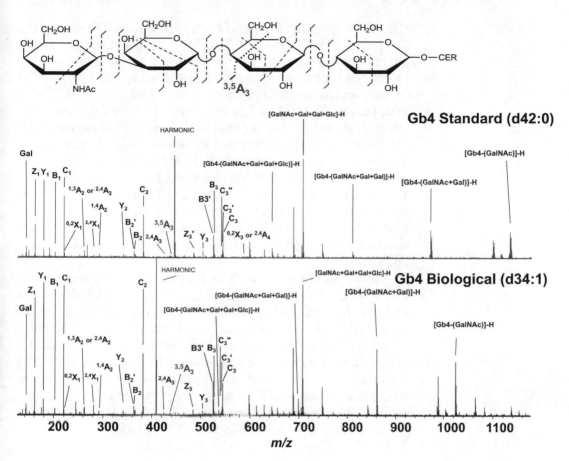

Fig. 5 Electron-Induced Dissociation (EID) MS/MS Verification of Gb4 Ganglioside. *Upper panel*: Cross ring cleavages in the sugar head group generated by the EID fragmentation. The $^{3,5}A_3$ fragment is the definitive fragment needed to determine that the linkage is 1–4 (Gb4) and not 1–3 (iGb4). *Upper Spectrum*: EID fragmentation spectrum of Gb4 (d42:0) standard. *Lower Spectrum*: EID fragmentation validation of Gb4 (d34:1) ganglioside in U373 glioma cells. Modified Bligh/Dyer aqueous phase biological extract from U373 glioma cells separated by nano-LC over a phenyl-hexal 50 μm i.d. × 8 cm column. Column eluent ionized by negative ion micro-ESI interfaced to a Bruker 9.4 T FT-ICR MS. Parent ion was "peak parked" at 25 nL/min. Spectrum shows EID fragmentation spectrum of the parent ion with the "diagnostic" $^{3,5}A_3$ fragment (s/n ~10:1) verifying a 1–4 linkage between the two galactose residues in the sugar head group, thus verifying that the biological lipid was in fact Gb4 and not iGb4

spectrum is shown in the middle panel of Fig. 5. The upper panel shows the cross ring cleavages generated by the EID fragmentation. The $^{3,5}A_3$ fragment in the sugar head group is the definitive fragment needed to determine that the linkage is 1–4 (Gb4) and not 1–3 (iGb4). The bottom panel shows the spectrum of the biological lipid sample extracted from the glioma cell lines. The aqueous phase of the modified Bligh/Dyer extraction from the glioma cell line was dried down and reconstituted in 80% methanol/ H₂O with the addition of 10 mM NH₄OAc and separated by nano-LC (Eksigent 1D system) in an 50 μm i.d. ×8 cm column

(New Objective) with self-packed phenyl-hexyl resin (Phenomenex). The gradient was flowed at 400 nL/min, and the eluent was flowed directly from the column into the micro-ESI source in negative ion mode. When the mass corresponding to the biological Gb4/iGb4 (d34:1) started to elute, the nano-LC was "peak parked" at 25 nL/min. Peak parking allowed plenty of time to isolate the ion in the external quadrupole and pass to the ICR cell for the accurate mass determination of the parent and subsequent EID fragmentation. The bottom spectrum shows the full EID spectrum of the biological with the characteristic $^{3,5}A_3$ fragment (s/n ~10:1), verifying a 1–4 linkage between the two galactose residues in the sugar head group and thus, confirming that the biological lipid was in fact Gb4 and not iGb4. The above example illustrates the power of the unique MS/MS capabilities of FT-ICR MS and their application to lipidomics.

Of course not every peak in a spectrum can be fragmented, thus once an ion has been identified by MS/MS a Kendrick Mass Scale can be used [68] to determine other ions that belong to the same lipid family based on exact mass of those ions. The Kendrick mass scale will be discussed in depth in the final section of this chapter. Even with all of the powers of ultrahigh-resolution FT-ICR MS and MS^n, there are still some lipid isomers that just cannot be distinguished. Ion mobility shows great promise to aid in the determination of these isomers, but to date there is no commercial vendor interfacing ion mobility to a FT-ICR MS.

4   Data Analysis

Data analysis in lipidomics is still in somewhat early stages, so many laboratories use customized, in-house tools for analysis of lipid data. However, increased interest in global lipidomics experiments has resulted in the development of commercial tools to facilitate data analysis (review: [69]). Most of these tools are not designed specifically for FT-ICR MS data, but they can be used for ultrahigh-resolution data.

For quantitative studies, commercial products designed for metabolomics, including Progenesis QI (Nonlinear Dynamics) and Elements (Proteome Software), can be used for lipidomic studies. SimLipid (Premier Biosoft International) is a commercial tool designed specifically for lipidomics. MS instrument vendors also have custom, proprietary software packages for the analysis of lipidomics data. Open source, lipid-specific software tools include ALEX ([70], designed specifically for data files from Thermo instruments) and LipidXplorer [71]. The open-source proteomics tool Skyline [72, 73] was recently adapted for use with lipid data [74]. For quantification, these tools rely on MS1-level extracted ion chromatograms of intact lipid species, and relative quantification is based upon a peak area ratio between the same species in different experimental conditions (normal vs. diseased, for example). Fragmentation spectra are used to assign the specific lipid species being quantified.

A critical component of both qualitative and quantitative studies is the identification of lipids. Tools for the identification of lipids use general metabolite databases such as the Human Metabolome Database (HMBD, [75, 76]), METLIN [77] and ChemSpider [78, 79], or lipid-specific databases such as Lipid MAPS Structure Database (LMSD, [80]) and LipidBlast [81]. Of these databases, HMDB is unique for its inclusion of fragmentation spectra from FT-ICR MS, but it is limited to human metabolites. Database searching is integrated into Progenesis QI and Elements, and the user can choose between several of the tools described above. In the case of SimLipid, identification is also integrated. However, relative quantification is based upon peak area relative to an added internal standard.

Assignment of lipid species in FT-ICR MS data is greatly facilitated by sub-ppm mass error. Due to the high resolving power and high mass accuracy, it is often possible to elucidate molecular formulas. However, this is not sufficient for the identification of isomeric species with the same molecular formula but different structures. In these cases, it is essential to obtain MS/MS (fragmentation) spectra.

The Kendrick mass scale [68] takes advantage of the high resolution of the FT-ICR MS and provides a method for extrapolation of structure for higher order lipid species. For this scale, the methylene group ($-CH_2-$) is arbitrarily assigned a value of 14.000 (erroneously attributed to the mass of nitrogen in Jones et al., 2004 [82]). Kendrick masses are calculated by multiplying the IUPAC mass by 14.000 and dividing by 14.01565 (the exact mass of a methylene group), and the Kendrick mass defect is calculated by subtracting the nominal Kendrick mass from the exact Kendrick mass [83]. The Kendrick mass scale was originally used in the field of petroleomics [83] and was later shown to have great utility in the analysis of lipidomics data [3]. Due to similarities in heteroatom composition, lipids that fall into the same functional class have the same Kendrick mass defect. Therefore, for a series of measured mass values, when a plot is generated with nominal Kendrick mass on the x-axis and the corresponding Kendrick mass defect on the y-axis, species in the same class fall on a horizontal line. Within these species, those that fall within a series based upon increasing numbers of methylene groups will differ in mass by a multiple of 14. Species that differ by a degree of unsaturation (i.e., a ring or a double bond) will differ in mass by two. Researchers can take advantage of these known patterns to map proposed identifications on lipid species for which a MS/MS spectrum has not been obtained.

5 Conclusions

Ultrahigh-resolution FT-ICR MS has several advantages for lipid identification: unparalleled accurate mass, ultrahigh-resolution, isotopic fine structure, and multiple unique MS/MS fragmentation capabilities. That being said, there are also shortcomings of the technique. As with high-resolution Orbitrap analysis, scan times tend to be slow; the higher the resolution, the slower the scan times. FT-ICR MS scan times can be decreased by going to higher magnetic fields and have also decreased with the introduction of the dynamically harmonized ICR cell. The slower scan times result in a reduction of the number of MS/MS scans that can be performed in an LC time scale. Capitalizing on sub-ppm accurate mass and using the Kendrick Mass Scale analysis compensates for the smaller number of MS/MS scans in a chromatographic run of biological lipids. There is also a large maintenance issue with an FT-ICR MS. The researcher needs to understand the care and maintenance of a high-field superconducting magnet. Finally, an experienced mass spectrometrist trained in FT-ICR MS is needed to efficiently extract the potential of this high-end analytical instrument. Ultrahigh-resolution FT-ICR MS is truly unequalled in the absolute determination of biological lipids and there are new capabilities being developed on these instruments all the time. The best scenario would be to have access to not only ultrahigh-resolution FT-ICR MS, but also utilize complementary lower resolution instruments in your lipid research as well.

Acknowledgments

The authors would like to acknowledge Huan He and Alan G. Marshall, at the NSF Funded FT-ICR Facility at the NHMFL, Tallahassee, FL for access to all the instruments in that facility and acquisition and analysis of the data in Figs. 1, 3, and 4. Thanks are given to Franklin E. Leach and Jonathan Amster at University of Georgia–Athens, Athens, GA for access to their 9.4 T FT-ICR MS and expertise in acquiring the data for Fig. 5 (Huan He was also crucial to this experiment). Finally, the authors would like to acknowledge Brooke L. Barnette and Shinji K. Strain at University of Texas Medical Branch-Galveston, Galveston, TX, for the helpful review of this manuscript.

References

1. Lydic TA, Busik JV, Reid GE (2014) A monophasic extraction strategy for the simultaneous lipidome analysis of polar and nonpolar retina lipids. J Lipid Res 55(8):1797–1809

2. Wood PL (2012) Lipidomics of Alzheimer's disease: current status. Alzheimers Res Ther 4(1):5

3. He H et al (2011) High mass accuracy and resolution facilitate identification of glyco-

sphingolipids and phospholipids. Int J Mass Spectrom 305(2–3):116–119

4. Fahy E et al (2005) A comprehensive classification system for lipids. J Lipid Res 46(5): 839–861

5. Park H, Zhou Y, Costello CE (2014) Direct analysis of sialylated or sulfated glycosphingolipids and other polar and neutral lipids using TLC-MS interfaces. J Lipid Res 55(4):773–781

6. Soares AF, Lei H, Gruetter R (2015) Characterization of hepatic fatty acids in mice with reduced liver fat by ultra-short echo time 1H-MRS at 14.1 T in vivo. NMR Biomed 28(8):1009–1020

7. Bojko B et al (2014) Solid-phase microextraction in metabolomics. Trends Anal Chem 61:168–180

8. He H et al (2007) Method for lipidomic analysis: p53 expression modulation of sulfatide, ganglioside, and phospholipid composition of U87 MG glioblastoma cells. Anal Chem 79(22): 8423–8430

9. Wood PL et al (2015) Non-targeted lipidomics of CSF and frontal cortex grey and white matter in control, mild cognitive impairment, and Alzheimer's disease subjects. Acta Neuropsychiatr 27(5):270–278

10. Abbassi-Ghadi N et al (2016) A comparison of DESI-MS and LC-MS for the lipidomic profiling of human cancer tissue. J Am Soc Mass Spectrom 27(2):255–264

11. Colsch B et al (2015) Lipidomic analysis of cerebrospinal fluid by mass spectrometry-based methods. J Inherit Metab Dis 38(1):53–64

12. Milman BL et al (2015) Comparative determination of fatty acid composition of low-molecular components of blood plasma by three mass spectrometry techniques: the 'old-new' exercise in lipidomics. J Anal Chem 70(14): 1601–1613

13. Lydic TA et al (2015) Rapid and comprehensive 'shotgun' lipidome profiling of colorectal cancer cell derived exosomes. Methods 87:83–95

14. Granafei S et al (2015) Identification of isobaric lyso-phosphatidylcholines in lipid extracts of gilthead sea bream (*Sparus aurata*) fillets by hydrophilic interaction liquid chromatography coupled to high-resolution Fourier-transform mass spectrometry. Anal Bioanal Chem 407(21):6391–6404

15. Martano C et al (2015) Rapid high performance liquid chromatography-high resolution mass spectrometry methodology for multiple prenol lipids analysis in zebrafish embryos. J Chromatogr A 1412:59–66

16. Bromke MA et al (2015) Liquid chromatography high-resolution mass spectrometry for fatty acid profiling. Plant J 81(3):529–536

17. He H et al (2011) Algae polar lipids characterized by online liquid chromatography coupled with hybrid linear quadrupole ion trap/Fourier transform ion cyclotron resonance mass spectrometry. Energy Fuel 25(10):4770–4775

18. Sudasinghe N et al (2015) Temperature-dependent lipid conversion and nonlipid composition of microalgal hydrothermal liquefaction oils monitored by Fourier transform ion cyclotron resonance mass spectrometry. Bioenergy Res 8(4):1962–1972

19. Serricchio M et al (2015) Flagellar membranes are rich in raft-forming phospholipids. Biol Open 4(9):1143–1153

20. Nazari M, Muddiman DC (2015) Cellular-level mass spectrometry imaging using infrared matrix-assisted laser desorption electrospray ionization (IR-MALDESI) by oversampling. Anal Bioanal Chem 407(8):2265–2271

21. Reis A et al (2013) A comparison of five lipid extraction solvent systems for lipidomic studies of human LDL. J Lipid Res 54(7):1812–1824

22. Wasntha T, Rupasinghe T (2013) Lipidomics: extraction protocols for biological matrices. In: Roessner U, Dias AD (eds) Metabolomics tools for natural product discovery: methods and protocols. Humana Press, Totowa, NJ, pp 71–80

23. Raterink R-J et al (2014) Recent developments in sample-pretreatment techniques for mass spectrometry-based metabolomics. Trends Anal Chem 61:157–167

24. Bligh EG, Dyer WJ (1959) A rapid method of total lipid extraction and purification. Can J Biochem Physiol 37(8):911–917

25. Folch J, Lees M, Sloane Stanley GH (1957) A simple method for the isolation and purification of total lipides from animal tissues. J Biol Chem 226(1):497–509

26. Matyash V et al (2008) Lipid extraction by methyl-tert-butyl ether for high-throughput lipidomics. J Lipid Res 49(5):1137–1146

27. Cajka T, Fiehn O (2014) Comprehensive analysis of lipids in biological systems by liquid chromatography-mass spectrometry. Trends Anal Chem 61:192–206

28. Zhao YY et al (2014) Ultra-performance liquid chromatography-mass spectrometry as a sensitive and powerful technology in lipidomic applications. Chem Biol Interact 220: 181–192

29. Uchikata T et al (2012) High-throughput phospholipid profiling system based on supercritical fluid extraction-supercritical fluid chromatography/mass spectrometry for dried plasma spot analysis. J Chromatogr A 1250:69–75

30. Williams JR, Clifford AA, al-Saidi SH (2002) Supercritical fluids and their applications in

biotechnology and related areas. Mol Biotechnol 22(3):263–286

31. Karas M, Hillenkamp F (1988) Laser desorption ionization of proteins with molecular masses exceeding 10000 daltons. Anal Chem 60(20):2299–2301

32. Tanaka K et al (1988) Protein and polymer analyses up to m/z 100 000 by laser ionization time-of-flight mass spectrometry. Rapid Commun Mass Spectrom 2(8):151–153

33. Caprioli RM, Farmer TB, Gile J (1997) Molecular imaging of biological samples: localization of peptides and proteins using MALDI-TOF MS. Anal Chem 69(23):4751–4760

34. Juhasz P, Costello CE (1992) Matrix-assisted laser desorption ionization time-of-flight mass spectrometry of underivatized and permethylated gangliosides. J Am Soc Mass Spectrom 3(8):785–796

35. Whitehouse CM et al (1985) Electrospray interface for liquid chromatographs and mass spectrometers. Anal Chem 57(3):675–679

36. Emmett MR, Caprioli RM (1994) Micro-electrospray mass spectrometry: ultra-high-sensitivity analysis of peptides and proteins. J Am Soc Mass Spectrom 5(7):605–613

37. Wilm MS, Mann M (1994) Electrospray and Taylor-Cone theory, Dole's beam of macromolecules at last? Int J Mass Spectrom Ion Process 136(2–3):167–180

38. Hendrickson CL, Emmett MR (1999) Electrospray ionization Fourier transform ion cyclotron resonance mass spectrometry. Annu Rev Phys Chem 50:517–536

39. Sparkman OD (2000) Mass spectrometry desk reference. Global View Publishing, p. 8, ISBN: 0-9660813-2-3.

40. Carroll DI et al (1974) Subpicogram detection system for gas-phase analysis based upon atmospheric-pressure ionization (Api) mass-spectrometry. Anal Chem 46(6):706–710

41. Henlon JD, Thomson BA, Dawson PH (1982) Determination of sulfa drugs in biological-fluids by liquid-chromatography mass-spectrometry mass-spectrometry. Anal Chem 54(3):451–456

42. Baim MA, Eatherton RL, Hill HH (1983) Ion mobility detector for gas chromatography with a direct photoionization source. Anal Chem 55(11):1761–1766

43. Syage JA, Short LC, Cai SS (2008) Atmospheric pressure photoionization—the second source for LC-MS? LCGC North Am 26(3):286–296

44. Imbert L et al (2012) Comparison of electrospray ionization, atmospheric pressure chemical ionization and atmospheric pressure photoionization for a lipidomic analysis of Leishmania donovani. J Chromatogr A 1242:75–83

45. Takats Z et al (2004) Mass spectrometry sampling under ambient conditions with desorption electrospray ionization. Science 306(5695):471–473

46. Levery SB (2005) Glycosphingolipid structural analysis and glycosphingolipidomics. Methods Enzymol 405:300–369

47. Almeida R et al (2015) Comprehensive lipidome analysis by shotgun lipidomics on a hybrid quadrupole-orbitrap-linear ion trap mass spectrometer. J Am Soc Mass Spectrom 26(1):133–148

48. Casanovas A et al (2014) Shotgun lipidomic analysis of chemically sulfated sterols compromises analytical sensitivity: recommendation for large-scale global lipidome analysis. Eur J Lipid Sci Technol 116(12):1618–1620

49. Schuhmann K et al (2011) Bottom-up shotgun lipidomics by higher energy collisional dissociation on LTQ Orbitrap mass spectrometers. Anal Chem 83(14):5480–5487

50. Schwudke D et al (2011) Shotgun lipidomics on high resolution mass spectrometers. Cold Spring Harb Perspect Biol 3(9):a004614

51. Jessome LL, Volmer DA (2006) Ion suppression: a major concern in mass spectrometry. LCGC North Am 24:83–89

52. Hofstadler SA et al (1996) Analysis of single cells with capillary electrophoresis electrospray ionization Fourier transform ion cyclotron resonance mass spectrometry. Rapid Commun Mass Spectrom 10(8):919–922

53. Maxwell EJ, Chen DDY (2008) Twenty years of interface development for capillary electrophoresis-electrospray ionization-mass spectrometry. Anal Chim Acta 627(1):25–33

54. Molnárné Guricza L, Schrader W (2015) New separation approach for asphaltene investigation: argentation chromatography coupled with ultrahigh-resolution mass spectrometry. Energy Fuel 29(10):6224–6230

55. Emmett MR et al (2006) Supercritical fluid chromatography reduction of hydrogen/deuterium back exchange in solution-phase hydrogen/deuterium exchange with mass spectrometric analysis. Anal Chem 78(19):7058–7060

56. Bamba T et al (2012) Metabolic profiling of lipids by supercritical fluid chromatography/mass spectrometry. J Chromatogr A 1250:212–219

57. Bamba T et al (2008) High throughput and exhaustive analysis of diverse lipids by using supercritical fluid chromatography-mass spectrometry for metabolomics. J Biosci Bioeng 105(5):460–469

58. Lee JW et al (2013) Simultaneous profiling of polar lipids by supercritical fluid chromatography/tandem mass spectrometry with methylation. J Chromatogr A 1279:98–107

59. Ashraf-Khorassani M et al (2015) Ultrahigh performance supercritical fluid chromatography of lipophilic compounds with application to synthetic and commercial biodiesel. J Chromatogr B Analyt Technol Biomed Life Sci 983:94–100

60. Comisaro MB, Marshall AG (1974) Fourier-transform ion-cyclotron resonance spectroscopy. Chem Phys Lett 25(2):282–283

61. Marshall AG, Hendrickson CL, Jackson GS (1998) Fourier transform ion cyclotron resonance mass spectrometry: a primer. Mass Spectrom Rev 17(1):1–35

62. Hendrickson CL et al (2015) 21 tesla Fourier transform ion cyclotron resonance mass spectrometer: a national resource for ultrahigh resolution mass analysis. J Am Soc Mass Spectrom 26(9):1626–1632

63. Shi SD-H, Hendrickson CL, Marshall AG (1998) Counting individual sulfur atoms in a protein by ultrahighresolution Fourier transform ion cyclotron resonance mass spectrometry: experimental resolution of isotopic fine structure in proteins. Proc Natl Acad Sci 95(20):11532–11537

64. Emmett MR et al (2014) Integrative biological analysis for neuropsychopharmacology. Neuropsychopharmacology 39(1):5–23

65. McFarland MA et al (2005) Structural characterization of the GM1 ganglioside by infrared multiphoton dissociation, electron capture dissociation, and electron detachment dissociation electrospray ionization FT-ICR MS/MS. J Am Soc Mass Spectrom 16(5):752–762

66. Kroes RA et al (2010) Overexpression of ST6GalNAcV, a ganglioside-specific alpha2,6-sialyltransferase, inhibits glioma growth in vivo. Proc Natl Acad Sci U S A 107(28):12646–12651

67. Wolff JJ et al (2008) Electron-induced dissociation of glycosaminoglycan tetrasaccharides. J Am Soc Mass Spectrom 19(10):1449–1458

68. Kendrick E (1963) A mass scale based on Ch2=14.0000 for high resolution mass spectrometry of organic compounds. Anal Chem 35(13):2146–2154

69. Hartler J et al (2013) Bioinformatics tools and challenges in structural analysis of lipidomics MS/MS data. Brief Bioinform 14(3):375–390

70. Husen P et al (2013) Analysis of lipid experiments (ALEX): a software framework for analysis of high-resolution shotgun lipidomics data. PLoS One 8(11):e79736

71. Herzog R et al (2012) LipidXplorer: a software for consensual cross-platform lipidomics. PLoS One 7(1):e29851

72. MacLean B et al (2010) Skyline: an open source document editor for creating and analyzing targeted proteomics experiments. Bioinformatics 26(7):966–968

73. Schilling B et al (2012) Platform-independent and label-free quantitation of proteomic data using MS1 extracted ion chromatograms in skyline: application to protein acetylation and phosphorylation. Mol Cell Proteomics 11(5):202–214

74. Peng B, Ahrends R (2016) Adaptation of skyline for targeted lipidomics. J Proteome Res 15(1):291–301

75. Wishart DS et al (2013) HMDB 3.0–the human metabolome database in 2013. Nucleic Acids Res 41(Database issue):D801–D807

76. Wishart DS et al (2007) HMDB: the human metabolome database. Nucleic Acids Res 35(Database issue):D521–D526

77. Smith CA et al (2005) METLIN: a metabolite mass spectral database. Ther Drug Monit 27(6):747–751

78. Little JL et al (2012) Identification of "known unknowns" utilizing accurate mass data and chemspider. J Am Soc Mass Spectrom 23(1):179–185

79. Pence HE, Williams A (2010) ChemSpider: an online chemical information resource. J Chem Educ 87(11):1123–1124

80. Sud M et al (2007) LMSD: LIPID MAPS structure database. Nucleic Acids Res 35:D527–D532

81. Kind T et al (2013) LipidBlast in silico tandem mass spectrometry database for lipid identification. Nat Methods 10(8):755–758

82. Jones JJ et al (2004) Strategies and data analysis techniques for lipid and phospholipid chemistry elucidation by intact cell MALDI-FTMS. J Am Soc Mass Spectrom 15(11):1665–1674

83. Hughey CA et al (2001) Kendrick mass defect spectrum: a compact visual analysis for ultrahigh-resolution broadband mass spectra. Anal Chem 73(19):4676–4681

Chapter 4

MALDI-Imaging Mass Spectrometry of Brain Lipids

Norelle C. Wildburger

Abstract

Advances in mass spectrometry (MS) over the past two decades have made MS an invaluable tool for the detection and characterization of biomolecules. One such biomolecule, lipids, is an important, but often times challenging species to analyze. Here, we describe the matrix-assisted laser desorption/ionization imaging mass spectrometry (MALDI-IMS) of lipids and glycolipids from mammalian brain in an untargeted and high-throughput approach.

Key words MALDI-IMS, brain, Shotgun lipidomics, Cryo-sectioning, Sublimation, ITO slides, Mass spectrometry

1 Introduction

Lipids are unique biomolecules with dual structural and function roles in biological systems. Lipids are crucial to the maintenance of cellular structure through the plasma membrane and critical determinants of membrane fluidity, which impacts cell signaling microdomains (e.g., lipid rafts or glycosynapses) [1] and membrane permeability. Functionally, lipids may serve as signal transduction molecules derived from cleavages or remodeling of structural membrane lipids (e.g., DAG synthesis via enzymatic cleavage of PIP2 or dephosphorylation of phosphatidic acid; Kennedy pathway) [2, 3].

The multifaceted and diverse roles of lipids in biological systems highlight the importance of analyzing and monitoring lipids and their metabolism. For instance, alterations in lipid metabolism frequently accompany diseases and disorders such as cancer, Alzheimer's disease (AD), and schizophrenia [4–9] in addition to well-characterized changes in proteins (e.g., amyloid-beta and tau in AD). This becomes imperative in the brain, which contains the highest lipid content compared to any other organ in the body and which has subregions that can be defined by a specific lipid type.

Paul Wood (ed.), *Lipidomics*, Neuromethods, vol. 125,
DOI 10.1007/978-1-4939-6946-3_4, © Springer Science+Business Media LLC 2017

Yet, lipidomic analysis has lagged behind proteomic analysis. The challenge lies in the fact that lipids are hydrophobic (e.g., ceramides) or amphipathic in nature—containing both hydrophobic and hydrophilic regions. Even the more "hydrophilic" glycolipids containing a polar sugar head group [10, 11] are challenging due to their hydrophobic hydrocarbon chains. This makes traditional nLC-MS/MS (with resin-based separations) challenging and impractical for many lipid classes. Compounding this is the high structural similarity among lipid classes and lack of database or de novo search mechanisms for their rapid identification creating an analytical bottleneck.

However, lipid analysis has been particularly successful with matrix-assisted laser desorption/ionization (MALDI). With advances in mass spectrometry technology, it is now possible to couple lipidomic profiling with in situ localization in biological samples, circumventing the need for lengthy extractions, which can result in significant sample loss, tedious derivation steps, lipid class biases, and loss of spatial organization information. The result is the generation of molecular–histological maps from the localization and identification of lipid biomolecules based on mass-to-charge ratio (m/z).

Here, we detail a MALDI-IMS protocol for untargeted or shotgun lipidomics of up to 200 lipids in brain tissue when no *a priori* knowledge is available (i.e., hypothesis-generating) in a given sample set comparing, for example, control *vs* disease or treated *vs* untreated. MALDI-IMS is followed by the confirmation of select *m/z* assignments by MS/MS, which yields chemical information used to derive lipid structure. We also emphasize throughout this protocol rigor in sample preparation and data acquisition for optimized downstream data analysis. To further support lipid findings from MALDI-IMS and MS/MS experiments, orthogonal validation with ESI-MS/MS is recommended; *see* Chap. 2 "High-Resolution Lipidomics with the Orbitrap" and Chap. 3 "High-Resolution Lipidomics with FT-ICR."

2   Materials

Prepare all solutions using LC-MS grade water (J.T. Baker) and analytical grade reagents. Prepare and store all reagents at room temperature unless indicated otherwise. Diligently follow all waste disposal regulations when disposing waste materials. Dispose of all biological material as biological waste.

2.1 Animals and Dissection Components

Follow all institutional, state, and federal regulations on the care and use of research animals.

1. Anesthesia: ketamine/xylazine solution (200 mg of ketamine and 20 mg of xylazine in 17 mL of saline) for mice [12] and isofluorane for rats [13].

2. Rodent and small animal guillotine for decapitation, if performing analysis on rats.

3. Timer.

4. Dewar of liquid nitrogen: Approximately 1 liter is recommended for processing four to six animals.

5. Solid metal block: We typically use a metal heating block inverted, so the flat side faces up.

6. Aluminum foil.

7. Styrofoam box: Use a thick-walled Styrofoam box that will fit the metal block. Liquid nitrogen will also be placed in the Styrofoam box, but not so much as to submerge the metal block (Fig. 1).

8. Glass or plastic petri dishes.

9. Conical tubes and tube rack: We use 50 mL conical tubes for rats and 15 mL conical tubes for mice for storing brain material at −80 °C.

10. Cryostat or access to a cryostat capable of making 10 μm slices consistently.

11. Razor blades.

12. Optional: Artists brush for small sections

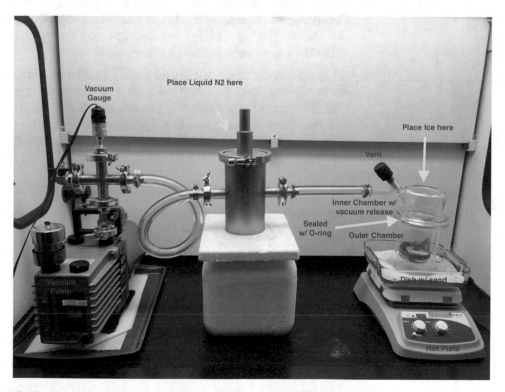

Fig. 1 Sublimation apparatus setup contained within a chemical fume hood

13. Conductive indium tin oxide (ITO)-coated glass slides (#237001; Bruker Daltonics, Bremen, Germany) (*see* **Note 1**).

14. Microscope slide box.

15. −80 °C freezer.

2.2 MALDI-IMS Sample Preparation Components

2.2.1 Follow all MSDS Safety Information for Chemicals and Reagents

1. LC-MS grade acetonitrile (J.T. Baker, Philipsburg, NJ).

2. LC-MS grade water (J.T. Baker, Philipsburg, NJ).

3. TFA (Pierce, Rockford, IL): Use to make a 0.1% (v/v) aqueous TFA solvent with LC-MS grade water.

4. Sonicating water bath.

5. Peptide calibration standard I (#206195; Bruker Daltonics, Bremen, Germany): Dissolve in 125 μL 0.1% TFA in LC-MS grade water. The various peptides and their *m/z* values are listed in Table 1. Make 3 μL aliquots in Eppendorf Lobind microcentrifuge tubes to prevent polymer contamination and sample loss. Store at −80 °C for up to 2 years.

6. 50 mM ammonium acetate solution: Weigh 192.7 mg of highest purity ammonium acetate (Sigma Chemical Company, St. Louis, MO, USA) and add LC-MS grade water to a total volume of 50 mL (*see* **Note 2**). For best results make fresh daily and chill at 2–8 °C.

7. 2,5-dihydroxybenzoic acid (DHB) (Sigma Chemical Company, St. Louis, MO, USA) (*see* **Note 3**): Weigh out 500 mg of DHB in a 15 mL conical tube and keep sealed and in the dark until use; DHB is light sensitive.

8. α-Cyano-4-hydroxycinnamic acid (HCCA) (Bruker Daltonics, Bremen, Germany) (*see* **Note 4**): HCCA power is stored at −20 °C. Matrix [approximately the amount that fits on the tip of a micro-spatula] is placed in a Eppendorf Lobind microcentrifuge tube and resuspended in 14 μL 0.1% TFA solvent and

Table 1
Peptide calibration standards

Peptide	Monoisotopic [M + H]$^+$	Average [M + H]$^+$
Angiotensin II	1046.5418	1047.19
Angiotensin I	1296.6848	1297.49
Substance P	1347.7354	1348.64
Bombesin	1619.8223	1620.86
ACTH clip 1-17	2093.0862	2094.43
ACTH 18-39	2465.1983	2466.68
Somatostatin	3147.471	3149.57

7 µL acetonitrile and sonicated in a water bath for 15 min. This process produces a super-saturated solution of HCCA. Next, briefly spin the tube on a desktop centrifuge to pellet the bulk of the matrix. An ideal supernatant solution will still be somewhat opaque. The supernatant is used in a 50/50 mixture with the calibration standard. Matrix is made fresh daily.

9. MALDI target holder; MTP TLC Adapter (#255595; Bruker Daltonics, Bremen, Germany) for 75 × 50 mm glass slides or MTP Slide Adapter II (#235380; Bruker Daltonics, Bremen, Germany) for two 75 × 25 × 0.9 mm glass slides.

10. Sublimation apparatus (Fig. 2) (*see* **Note 5**): Sublimation apparatus consists of the following components.

    (a) Hot plate.

    (b) Glass dish with sand.

    (c) Cable temperature measuring transducer.

    (d) Vacuum pump.

    (e) Vacuum gauge.

    (f) Cold trap.

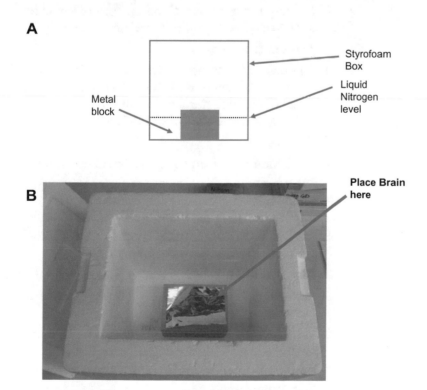

Fig. 2 Schematic of liquid nitrogen vapor freezing. (**a**) Diagram indicating appropriate liquid nitrogen levels relative to metal block. (**b**) Image of real liquid nitrogen vapor freezing setup

(g) Ice.

(h) Liquid nitrogen.

(i) Outer vessel.

(j) Inner vessel.

11. Copper tape.

12. Vacuum desiccator for sample dehydration and drying.

13. Desiccant box for sample storage.

14. Black poster board cut to match the shape and dimensions of the MTP Slide Adapter II when locked.

2.3 Optical Imaging Supplies

1. Glass petri dish.

2. Hematoxylin.

3. Eosin.

4. Water: Ultrapure water (purified deionized water to attain a sensitivity of 18 MΩ cm at 25 °C).

5. Ethanol: Graded series of 100%, 95%, and 70% made with ultrapure water.

6. Xylene: All the steps using xylene must be performed in a chemical fume hood.

7. Staining jars or dishes: Coplin jar or otherwise for graded ethanol series, hematoxylin, eosin, and xylene.

8. Light microscope.

9. Optional: Microscope slides (Superfrost™ Plus). Use only if you want to stain (e.g., H&E) serial adjacent sections of brain tissue rather than the slices taken for MALDI-IMS (i.e., H&E staining of tissue after MALDI-IMS).

10. Coverslips.

11. DPX Mounting media (Sigma Chemical Company, St. Louis, MO, USA).

12. Slide box.

2.4 Analytical Equipment and Supplies

1. MALDI-TOF/TOF mass spectrometer (Ultraflex extreme, Bruker Daltonics, Bremen, Germany).

2. FlexControl 3.4 software (Bruker Daltonics, Bremen, Germany).

3. FlexImaging 3.0 software (Bruker Daltonics, Bremen, Germany).

4. FlexAnalysis 3.0 software (Bruker Daltonics, Bremen, Germany).

5. Computer to operate MALDI-TOF/TOF with high-resolution scanner attached.

3 Methods

Carry out all the procedures at room temperature unless otherwise specified.

3.1 Tissue Dissection and Sectioning

Mice: anesthetize animals by intraperitoneal injection of ketamine/xylazine solution (200 mg of ketamine and 20 mg of xylazine in 17 mL of saline) at a dosage of 0.15 mg/10 g body weight and sacrifice by CO_2 inhalation [12]. Rats: anesthetize with isofluorane followed by decapitation with the rodent and small animal guillotine [13].

1. On a glass petri dish, rapidly remove the brain in under 3 min. Use the timer to ensure you are within this range to prevent lipid alterations due to prolonged exposed at room temperature. Alternatively, the dissection petri dish can be placed on ice for the procedure, particularly if the duration is increased due to micro-dissection of brain subregions.

2. Place the dissected brain on the center of a square piece of aluminum foil and place on the metal block inside the Styrofoam box with liquid nitrogen (the liquid nitrogen should not submerge the block) (see Fig. 2). Close the lid of the Styrofoam box and set the time to 2 min. The liquid nitrogen vapor will rapidly freeze the brain tissue. Frozen tissue will be paler in color than unfrozen tissue. Avoid liquid nitrogen touching the tissue, as it will cause the sample to fracture and break.

3. Once the tissue is completely frozen remove aluminum foil from the Styrofoam box and place the tissue in a conical tube and keep in the dewar of liquid nitrogen until finished with all dissections.

4. After dissections are complete, transfer the conical tubes from liquid nitrogen dewar to a −80 °C freezer for storage until sectioning (see **Note 6**).

5. Ensure the cyrostat is set to −20 °C.

6. For sectioning tissue, remove a tube with sample from the -80°C and equilibrate for 2 h in the cryostat to −20 °C.

7. Take ITO slides and with the correct side (ITO side, see **Note 1**) facing up, use a silver sharpie mark three to four corners of the slide with X's, T's, and/or L's. This ensures 1) the slide is correctly oriented, but 2) more importantly, will serve as reference points for teaching the MALDI software at the sample position. **Important:** Keep these marking close to the edges of the slide.

8. Place in slides in the cryostat to chill.

9. Next, with a razor blade carefully cut a flat edge in the brain. For example, for sectioning rostral to caudal, cut at

the cerebellum to create a flat surface to mount the brain on the sample holder with a droplet of water. Once the water freezes and the sample is stable, add additional droplets of water around the cerebellum to secure the brain.

10. Mount the sample holder onto the chuck and use the cryostat to make 10 μm slices (*see* **Note 7**) [14].

11. Carefully pick up the tissue slice with a brush and place on an ITO slide (keeping it within the cryostat). Use the brush to unfold the tissue. When ready carefully remove the slide from the cryostat and warm to room temperature to thaw-mount the tissue (*see* **Note 8**). Be mindful of changes in tissue orientation when transferring to ITO slide.

12. Section and mount all biological samples and their technical replicates in a block randomized fashion (*see*, Oberg and Vitek, 2009 [15]).

13. Place all ITO slides with tissue samples in the slide box and wrap with saran wrap to prevent frost buildup on tissue slides during -80 °C freezer storage.

14. Store in −80 °C until MALDI-IMS analysis. Samples can be stored in −80 °C for up to a year before detrimental effects on quality of MALDI-IMS data.

3.2 Sample Preparation for MALDI-IMS

1. Remove the sample slide to analyze from the slide box in the −80 °C freezer.

2. Let the slide equilibrate to room temperature (~15 min).

3. Dry the slice in the vacuum desiccator for 1 h.

4. Wash each tissue slice on the slide three times with chilled 50 mM ammonium acetate. Apply the solution, covering the entire tissue, and aspirate carefully with a pipette or by placing the slide on an incline and tapping onto a Kimwipe. Using vacuum aspiration may be deleterious to the slice. For one wash of a sagittal section of a mouse brain, 60 μL of 50 mM ammonium acetate is used.

5. Dry the slice under vacuum for 3 h.

6. Place the adapter upside-down on a table or lab bench and retract the beige colored retaining tabs on either side.

7. Place the slide into the adaptor with the tissue side facing down, gently lowering. Protruding screws on the front of the adapter prevent the tissue from contacting the surface.

8. Push the beige colored retaining tabs toward the center to lock the slide into place. The slide will be flush with the holder and the tissue to be imaged will face outward (i.e., the same side as adaptor name and product number).

9. Place the black poster board cutout behind the slide. This will enhance the image contrast when acquiring a JPEG or TIFF image in the scanner.

10. Place the adapter in the scanner facedown. Protruding screws on the front of the adapter prevent the tissue from contacting the surface.

11. Acquire a moderate resolution JPEG or TIFF image and save. **Important:** this must be done **before** matrix sublimation.

12. While the slide is still in the adapter, take a plastic lid from a 96-well plate (will numbered columns and lettered rows) and place over the adapter. Draw or transfer coordinates of the tissue on the slide to the plastic lid. This will aid in finding the sample inside the instrument. **Important:** this must be done **before** matrix sublimation.

13. Place the temperature sensing cable underneath the outer chamber of the sublimation apparatus and set the hotplate so that the temperature measuring transducer readout is 120 °C. With our setup, a hotplate setting of 186 °C produces 120 °C reading of the outer chamber.

14. Open inlet to the pump (#1), and close the outlet (#2) (*see* Fig. 2).

15. Using copper tape, tape the slide (without the adapter) to the bottom of the inner chamber of the sublimation apparatus so that the tissue faces downward into the outer chamber (Fig. 2). **Important:** place the copper tape around the periphery of the glass slide making sure to cover the reference points for teaching (i.e., "X's, T's, and/or L's"; *see* Subheading 3.1, **step 8**).

16. Place the DHB matrix in the bottom of the outer chamber and distribute evenly.

17. Place liquid nitrogen in the dewar of the cold trap and tighten into place.

18. Place the inner chamber into the outer chamber and attach tubing to the cold trap. Ensure that the O-ring is undamaged and in place.

19. Place ice in the opening atop the inner chamber. This is done last so that the sample does not build up condensation.

20. Turn on the vacuum and keep the vent partially open. If open too much the vacuum will not pump down.

21. Once the vacuum reaches 200 mTorr start the timer for 5 min.

22. After 5 min: close the vent, turn off the vacuum, close inlet to the pump (#1), and open the outlet (#2) (*see* Fig. 2).

23. Relieve the negative pressure by *slowly* opening the vent. Detach tubing from cold trap to inner chamber, remove the inner chamber, discard ice, and remove slide from the bottom of the inner chamber.

24. Taking a 3 μL aliquot of peptide calibration standard I, mix 50/50 (v/v) with prepared HCCA supernatant solution (Subheading 2.2.1, **item 7**), and place 1 μL spots in triplicate

on an un-sublimated portion of the slide. Use the silver sharpie to mark three spots next to the calibration standard spots to aid in localization within the MALDI.

25. Place the slide in the dessicant box to dry the calibration spots.

3.3 MALDI-IMS

The following steps describe MALD-IMS acquisition using a Ultraflex extreme MALDI-TOF/TOF mass spectrometer (Bruker Daltonics, Bremen, Germany). However, these steps may be generally applicable to other MALDI instruments and software.

It is the experimenter's responsibility to ensure the appropriate controls (positive and negative) are included in each sample set as well as the appropriate number of biological and technical replicates, based on power analysis, to derive statistically significant and meaningful conclusions.

1. Place slide in the adapter as described above (Subheading 3.2, **steps 6–8**).

2. Load the adaptor containing the slide into the MALDI.

3. Open both FlexImaging and FlexControl.

4. In FlexControl, create a new sequence, and define the raster width.

5. Upload the JGEP or TIFF image of the slide taken in Subheading 3.2, **step 11**.

6. Using the polygon tool, define your region-of-interest (ROI) on the tissue. This can be the whole tissue itself or a subregion within the tissue. A ROI defines the area which spectra will be acquired and processed. **Important:** the order you define your tissue ROIs is the order of data acquisition. Block randomize your selections to counter any instrumental drift that may skew sample results (e.g., if all normal tissues were run first, then all diseased tissues) [15].

7. Next teach the instrument the sample reference points made in Subheading 3.1, **step 8**. This primarily consists of switching to FlexControl, searching for the first prominent reference mark in the live video feed of your sample slide, adjusting the position so that it is under the cross-hairs (*see* **Note 9**) and switching back to FlexImaging and selecting the same mark (also under cross-hairs). Repeat this procedure two more times. A minimum of three reference points is required.

8. In FlexControl load the appropriate imaging method. In our studies, we use a mass range spanning $300–1600$ m/z and all lipid imaging is acquired in reflectron mode. Instrument settings vary based on instrument setup and selected mass range. These settings are more numerous than can be adequately described in this chapter; consult the instrument user manual for detailed instrument parameters and settings. Nonetheless, the most important parameters to optimize are acceleration

voltage (IS1), laser power, pulsed ion extraction (PIE), detector settings, and matrix suppression. These parameters should be optimized prior to sample data acquisition.

9. Move to the calibration standard spotted on the target. Use the silver markings as a guide to their location.

10. Calibrate the instrument using the cubic enhanced method starting at 200 ppm and successively decreasing to 20 ppm saving the method each time. Seven calibration standards are required. Reject any calibrant that deviates by ±10 ppm. If imaging low molecular weight species (i.e., < 500 *m/z*), it is advisable to include calibration on the DHB (matrix) ions to improve signal standard deviation in the low mass range. Adjust laser energy to maximize signal and resolution. **Important:** the instrument must be calibrated every time the adapter is inserted or removed from the instrument.

11. Move to a location on the edge of your tissue using the plate lid made in Subheading 3.2, **step 12** as a guide and optimize the laser power. Save the method once again.

12. Click autoexecute to start data acquisition (*see* **Note 10**).

13. Once complete, images can be displayed of lipids based on their *m/z* and site localization (Fig. 3).

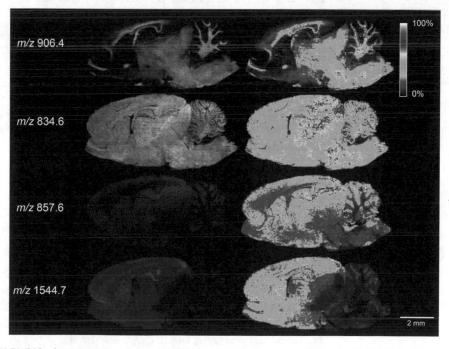

Fig. 3 MALDI-IMS of rat pup 17 days post-natal. Lipid ion images of sagittal sections of P17 rat pup (*left*) with corresponding heat maps of lipid ion intensity in situ *(right)*. *m/z's* 906.4, 834.6, 857.6, and 1544.7 (*top to bottom*). Note the regional differences in lipid localization. *m/z* 1544.7, a likely glycolipid due to its high mass, is absent in the caudal portion of the brain, while 906.4 is enriched in the white matter of the cerebellum (Arbor vitae) but not in the gray matter

3.4 MALDI-IMS Data Analysis

There are a number of tools available for the analysis of MALDI-IMS data. msIQuant [16] is a freely available software package developed at the Uppsala University. Limitations to this are primarily based on computational power of lab computers. Commercially available software includes those developed by ImaBiotech (Quantinetix™) and ClinProTools from Bruker Daltonics. Alternatively, imaging data may be normalized to the total ion current (recommended) and exported to Excel or R for analysis of lipid abundances with the statistical test appropriate for the experimental question at hand. All spectra should be processed with the same baseline subtraction and peak picking protocols. Lipid assignments, based on mass-to-charge ratio within a given tolerance (typically ±0.5 Da), are made through the LIPID Metabolites and Pathways Strategy Database (LipidMaps http://www.lipidmaps.org) [17, 18].

3.5 MS/MS Experiments

After data analysis in Subheading 3.4 is complete, confirm the identification of significantly differentially expressed lipid(s).

1. Prepare for MS/MS acquisition by calibrating the instrument as described in Subheading 3.3, **step 10**.

2. Collect spectra of interest in reflectron mode. Save and send to FlexAnalysis.

3. In FlexAnalysis, select the lipid of interest and add to "MS/MS list."

4. In FlexControl select the LIFT method and select your lipid *m/z* value.

5. Collect parent ion spectra.

6. Begin collection of fragment ions from the parent. Adjusting the laser power may be necessary. If the instrument is equipped with an external gas supply, use collision-induced dissociation (CID) fragmentation for best results.

7. Save the spectrum and manually annotate. Use only those fragment ion peaks with signal-to-noise (S/N) ratio > 3 (*see* **Notes 11** and **12**).

3.6 Optical Imaging (optional)

1. When MALDI-IMS acquisition is complete, remove adapter from instrument and remove the slide from the adapter.

2. In a glass petri dish pour 100% ethanol and gently lower the matrix covered slide into the dish and gently swirl. The matrix will dissolve and lift off the tissue (~15–30 s). If necessary repeat a second time to ensure all the matrix particles are removed. **Important:** tissue can easily be damaged in this step if care is not taken. A microscope can also be useful to observe complete matrix removal.

3. Using staining jars, place the slide in a series of 95% ethanol, 70% ethanol, and ultrapure water for 30 s each.

4. Stain with hematoxylin for 3 min.

5. Wash with the following for 30 s each: ultrapure water, 70% ethanol, and 95% ethanol.

6. Stain with eosin for 1 min.

7. Wash with the following for 30 s each: 95% ethanol and 100% ethanol.

8. Fix with xylene for 2 min. **Important:** this step must be performed in a fume hood.

9. Lay out to dry in the fume hood. Once dry, use mounting media to adhere coverslips.

10. Using a light microscope, image the tissue and/or ROI within the tissue.

11. Store slides in a slide box, kept at room temperature.

4   Notes

1. The ITO slides are light sensitive and must be kept in the dark. Also we have noticed in our experience that wrapping loose glass slides in Parafilm or any other similar material will leach polymer onto the slides and affect MS signal intensity and S/N. In addition, the slides are coated with ITO on only one side. Once receiving the slides, it is advisable to place a tick mark in permanent ink across the tops of the all slides within the box to demarcate the ITO-coated from uncoated side. Conductivity can be tested with a multimeter.

2. Through extensive testing 50 mM ammonium acetate was found to be the best wash buffer for brain tissue for a broad spectrum of lipids in an untargeted lipidomics approach based on overall signal intensity and S/N. However, this is not to say that for a more focused lipidomics approach (e.g., phospholipids, specifically) that alternative wash buffers would be more ideal.

3. Since the process of sublimation will purify DHB, low purity matrix may be purchased. Importantly, DHB does not undergo decomposition under sublimation conditions (i.e., reduced pressure and elevated temperature) [19].

4. HCCA is purchased at the highest purity available. The peptide calibration standard is mixed 50/50 (v/v) with the matrix and spotted on the ITO sample slide. HCCA is not sublimated onto the slide and thus not purified like DHB (*see* **Note 3**).

5. The glassware component of the sublimation apparatus in Fig. 1 was custom made, but all other components were purchased through scientific vendors. Sublimation is the choice method for applying matrix to the sample because (1) low-grade matrix, and thus less expensive matrix can be purchased (*see* **Note 4**), (2) minimal sample handling, and importantly (3) consistent matrix coverage across the tissue. Even deposition of matrix is crucial in MALDI-IMS as topological inconsistencies in matrix coverage will bias lipid signal intensity and S/N across different regions.

6. We use fresh frozen tissue so that we may perform imaging in both positive and negative ion modes. Most embedding resins (e.g., OCT) are rich in polymer components, which ionize readily and due to their abundance and ionization efficiency cause ion suppression making positive ion mode imaging impossible, but does not preclude negative ion mode. Some tissues, like spinal cord for instance, must be embedded in order to perform cryo-sectioning. However, gelatin and ice have proven to be MALDI-friendly embedding media [20].

7. Based on the work of Sugiura et al., we performed tissue thickness optimization for MALDI-IMS. We found that 10 μm slices for brain resulted in the best peak intensity and S/N [14]. It is important to optimize this prior finalizing the imaging protocol.

8. A simpler alternative is to invert a warm slide over the sold tissue sample and thaw-mount it to the surface of the warm slide. In this approach, a layer of ice remains on the cryostat sample platform. This may contain analytes of interest that may be lost in this process [20].

9. The purpose of using X's, T's, and/or L's as reference marks is that each of these letters has a sharp 90° or 45° angle that is especially prominent and ideal for teaching the instrument sample slide position. Move the cross-hair to these angles for optimal instrument teaching.

10. The length of data acquisition is primarily determined by the (1) the lateral resolution specified, (2) the size of the tissue or ROI to be image, and (3) number of samples per slide to be imaged. As the instrument will rapidly accumulate matrix debris, which may adversely affect sample results, it is advisable not to have long acquisition times.

11. Confirmatory identification can also be performed on a standard of the lipid species of interest for side-by-side comparison of the biological species and standard.

12. To complement MALDI-IMS and provide orthogonal validation, replicate biological samples may be processed by solvent: solvent extraction for targeted or untargeted lipidomics by ESI-MS/MS as previously described [12]. *See* Chap. 1 "High-Resolution Lipidomics with the Orbitrap" and Chap. 2 "High-Resolution Lipidomics with FT-ICR."

References

1. Hakomori S (2004) Glycosynapses: microdomains controlling carbohydrate-dependent cell adhesion and signaling. An Acad Bras Cienc 76(3):553–572

2. Berridge MJ (1987) Inositol trisphosphate and diacylglycerol: two interacting second messengers. Annu Rev Biochem 56:159–193

3. Kennedy EP (1987) Metabolism and function of membrane lipids. Klin Wochenschr 65(5):205–212

4. Santos CR, Schulze A (2012) Lipid metabolism in cancer. FEBS J 279(15):2610–2623

5. Wood PL (2012) Lipidomics of Alzheimer's disease: current status. Alzheimers Res Ther 4(1):5

6. Wood PL, Barnette BL, Kaye JA, Quinn JF, Woltjer RL (2015) Non-targeted lipidomics of CSF and frontal cortex grey and white matter in control, mild cognitive impairment, and Alzheimer's disease subjects. Acta Neuropsychiatr 27(5):270–278

7. Wood PL et al (2015) Targeted Lipidomics of Fontal Cortex and Plasma Diacylglycerols (DAG) in mild cognitive impairment and Alzheimer's disease: validation of DAG accumulation early in the pathophysiology of Alzheimer's disease. J Alzheimers Dis 48(2):537–546

8. Wood PL, Unfried G, Whitehead W, Phillipps A, Wood JA (2015) Dysfunctional plasmalogen dynamics in the plasma and platelets of patients with schizophrenia. Schizophr Res 161(2–3):506–510

9. Wood PL, Filiou MD, Otte DM, Zimmer A, Turck CW (2014) Lipidomics reveals dysfunctional glycosynapses in schizophrenia and the G72/G30 transgenic mouse. Schizophr Res 159(2–3):365–369

10. He H et al (2010) Polar lipid remodeling and increased sulfatide expression are associated with the glioma therapeutic candidates, wild type p53 elevation and the topoisomerase-1 inhibitor, irinotecan. Glycoconj J 27(1):27–38

11. He H et al (2007) Method for lipidomic analysis: p53 expression modulation of sulfatide, ganglioside, and phospholipid composition of U87 MG glioblastoma cells. Anal Chem 79(22):8423–8430

12. Wildburger NC et al (2015) ESI-MS/MS and MALDI-IMS localization reveal alterations in phosphatidic acid, diacylglycerol, and DHA in glioma stem cell xenografts. J Proteome Res 14(6):2511–2519

13. Shavkunov AS et al (2013) The fibroblast growth factor 14-voltage-gated sodium channel complex is a new target of glycogen synthase kinase 3 (GSK3). J Biol Chem 288(27):19370–19385

14. Sugiura Y, Shimma S, Setou M (2006) Thin sectioning improves the peak intensity and signal to-noise ratio in direct tissue mass spectrometry. J Mass Spectrom Soc Jpn 54(2):45–48

15. Oberg AL, Vitek O (2009) Statistical design of quantitative mass spectrometry-based proteomic experiments. J Proteome Res 8(5):2144–2156

16. Källback P, Shariatgorji M, Nilsson A, Andrén PE (2012) Novel mass spectrometry imaging software assisting labeled normalization and quantitation of drugs and neuropeptides directly in tissue sections. J Proteomics 75(16):4941–4951

17. Fahy E, Sud M, Cotter D, Subramaniam S (2007) LIPID MAPS online tools for lipid research. *Nucleic Acids Res* 35(Web Server issue):W606–W612

18. Fahy E et al (2009) Update of the LIPID MAPS comprehensive classification system for lipids. J Lipid Res 50(Suppl):S9–14

19. Hankin JA, Barkley RM, Murphy RC (2007) Sublimation as a method of matrix application for mass spectrometric imaging. J Am Soc Mass Spectrom 18(9):1646–1652

20. Norris JL, Caprioli RM (2013) Analysis of tissue specimens by matrix-assisted laser desorption/ionization imaging mass spectrometry in biological and clinical research. Chem Rev 113(4):2309–2342

Chapter 5

Ion-Mobility Mass Spectrometry for Lipidomics Applications

Giuseppe Paglia, Bindesh Shrestha, and Giuseppe Astarita

Abstract

Among lipidomics' major challenges, there is the molecular complexity of the lipidome. State-of-the-art technology, such as ion mobility (IM) spectrometry, is a promising new tool for supporting lipidomics research. IM is a gas-phase electrophoretic technique that enables the separation of ions in the gas phase according to their charge, shape, and size. IM separation, which occurs on a time scale of milliseconds, is compatible with modern mass spectrometers with microsecond scan speeds. Thus, IM-mass spectrometry (MS) can be integrated into conventional lipidomics MS workflows to improve separation and enhance peak capacity, spectral clarity, and fragmentation specificity. Furthermore, IM allows for the determination of the collision cross section (CCS), an orthogonal physicochemical measure that can be used, together with accurate mass and fragmentation information, to increase the confidence of lipid identification. In recent years, the expanding sophistication of hardware and software products has enabled IM-MS to perform an increasingly important role in traditional lipidomic approaches. In this chapter, we present IM-MS procedures for lipidomics research.

Key words IM, Lipid, Collision cross section, Drift time, Travelling-wave IM

1 Introduction

Lipid identification remains the most challenging step in a lipidomic workflow because of the chemical complexity of the lipidome [1–7]. In recent years, ion mobility (IM) technologies have supported lipidomic applications [8–16]. IM is a gas-phase electrophoretic technique that enables the separation of ions within an ion-mobility cell, a chamber filled with a buffer gas, like nitrogen, to which an electrical field is applied [17, 18]. The time required for lipid ions to migrate through the IM separation cell – the drift time – depends mostly on the frequency of collisional events between the ions and the buffer gas. Thus, drift time values are directly related to the shape, size, and charge of the lipid ions, as well as to the nature of the buffer gas (Fig. 1).

Paul Wood (ed.), *Lipidomics*, Neuromethods, vol. 125,
DOI 10.1007/978-1-4939-6946-3_5, © Springer Science+Business Media LLC 2017

The IM separation occurs on a timescale of milliseconds, making it suitable for coupling with mass spectrometry (MS), in which the detection usually occurs in microseconds. Three IM-MS technologies are currently commercially available: drift-time ion-mobility spectrometry (DTIMS), traveling-wave IM spectrometry (TWIMS), and differential-mobility spectrometry, the last of which is known also as field–asymmetric IM spectrometry (FAIMS). Comprehensive reviews addressing in detail the various IM techniques have been presented by Kanu et al. [17] and Lapthorn et al. [18]. By plotting mobility versus mass, it is possible to differentiate lipids from other classes of biomolecules such as peptides, carbohydrates, and oligosaccharides [19–21]. Lipid classes and subclasses fall into distinct trend lines on a m/z-mobility plot, facilitating the feature annotation of unknown lipid structures [11, 20, 22–25].

In recent years, exploiting both hardware and software advances; IM-MS has been increasingly used in traditional lipidomic approaches (Fig. 1). In this chapter, we provide experimental protocols on TWIM-MS from the perspective of our laboratory's experience in the field.

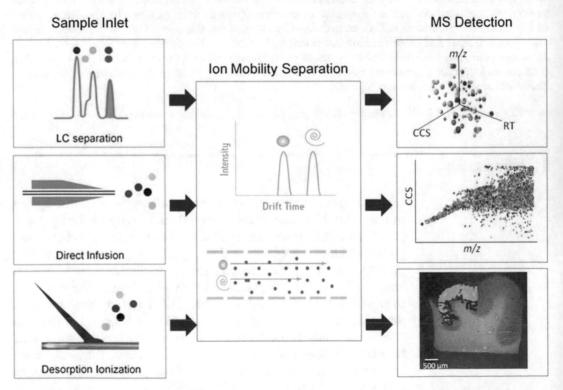

Fig. 1 Ion-mobility (IM) mass spectrometry (MS) can be used with traditional lipidomic approaches, such as chromatography, e.g., liquid chromatography (LC–IM–MS) and direct analysis. Independently of the inlet source and the ionization mode, lipid ions are separated before MS detection by their drift time, which is determined by their charge, size, and shape. Drift-time information can be converted to collision cross section (CCS), a measure of the shape of molecules. CCS provides an additional coordinate for identifying and increasing the signal-to-noise ratio

2 Materials

2.1 Equipment

1. Liquid chromatography system such as an ACQUITY UPLC® I-Class System (Waters Corporation, Milford, Massachusetts, USA) (*see* **Note 1**).

2. Desorption Electrospray Ionization (DESI) Source (Prosolia Inc., Indianapolis, Indiana, USA).

3. MS system: an IM time-of-flight mass spectrometer. For example, TWIMS-MS instruments including a SYNAPT High Definition Mass Spectrometry (HDMS).

4. Cryostat (e.g., CM3050 S, Leica Buffalo Grove, Illinois, USA).

2.2 Reagents

1. Natural lipids and lipid extracts (Avanti Polar Lipids, Alabaster, Alabama, USA).

2. Poly-DL-alanine (Sigma-Aldrich, St. Louis, Missouri, USA; cat # p9003).

3. CCS Major mix calibration solution (Waters, part number 186008113).

4. LCMS QCRM mix (Waters, part number 186006963).

5. Ion Mobility System Suitability LipidoMix™ Kit (Avanti Polar Lipids, cat # 791500).

6. Carboxymethylcellulose sodium salt (Sigma-Aldrich; cat # C4888).

7. Leucine encephalin (Sigma-Aldrich).

8. Sodium Formate (Sigma-Aldrich).

9. Acetic acid (Sigma-Aldrich).

10. Ammonium acetate (Sigma-Aldrich).

11. Solvents: isopropanol (IPA), methanol and acetonitrile (Thermo Fisher Scientific, Somerset, New Jersey, USA). Use only HPLC grade or better solvents.

2.3 Supplies

1. UPLC column: Charged Surface Hybrid (CSH™) C18 (2.1 × 100 mm) 1.7 μm (Waters).

2. Standard microscope glass slides (75 × 25 mm; e.g., VWR cat # 490013).

3. Glass tubes with caps including a Teflon-covered liner.

4. Gases. Routinely, IM measurements are made using nitrogen (N_2). Nevertheless, IM resolution can be optimized using other gases. The SYNAPT G2-Si and SYNAPT G2-S instruments can be used with 12 different IM gasses. Table 1 describes the expanded list of IM gases. In all cases, the gas purity must be at

Table 1
Supported IM gases on SYNAPT HDMS instruments

Nitrogen (N$_2$)	Neon (Ne)
Argon (Ar)	Nitrogen dioxide (NO$_2$)
Carbon Dioxide (CO$_2$)	Nitrogen oxide (NO)
Carbon Monoxide (CO)	Nitrous oxide (N$_2$O)
Helium (He)	Oxygen (O$_2$)
Hydrogen (H$_2$)	Sulfur hexafluoride (SF$_6$)

least 99.5%. Note that some gasses can require the use of specialized regulators, handling, or safety considerations.

2.4 Software

1. MassLynx (Waters) is used to acquire data for the SYNAPT HDMS system.

2. Progenesis® QI (Nonlinear Dynamics, Newcastle, UK) or UNIFI® (Waters) is used for processing and analyzing IM information for both qualitative and quantitative applications.

3. DriftScope™ software (Waters) is used to extract regions of interest for different molecules on the basis of selective drift-time extraction of mass spectra.

4. HDI® Imaging Software (Waters) is used for DESI experiments.

3 Methods

3.1 IM-MS Parameters: Setting and Optimization

1. Ensure IM settings are correct for your lipid samples. Prepare a fresh standard solution in isopropanol and use it for the optimization of IM settings. Several lipid extracts sample are commercially available from Avanti Polar. Select the lipid extracts that best fit your need and use it as system-suitability standards to confirm visualization and separation of the wide range of lipid classes present in biological samples. Use glass tubes with caps that include a Teflon-covered liner, to avoid leaching contaminants into organic solutions.

2. Directly infuse the relevant lipid mixture at the rate of 5 µL/min, adjusting IM parameters to control IM separation [26]. We suggest starting the optimization process by using the parameters presented in Tables 2 and 3. Then, ensure that all lipid masses fit in the IM separation window by adjusting the following parameters: IMS wave velocity and IMS wave Height cell. By increasing or decreasing these two parameters you can expand or narrow the drift-time distribution profile.

Table 2
Representative MS settings

Polarity	Flow rate (μL/min)	Cone Voltage (V)	Source Temperature (°C)	Desolvation Temperature (°C)	Desolvation Gas flow (L/h)	Cone Gas flow
ES+	400	30	110	450	450	20
ES−	400	40	110	500	500	20
Polarity	*EDC delay Coefficient*	*Optic Mode*	*MS scan rate*	*Lock mass Solution*	*Lock mass Flow rate (μL/min)*	
ES+	1.58	Resolution	0.2 Scan/s	Leucine encephalin (2 μg/mL)	15	
ES−	1.58	Resolution	0.2 Scan/s	Leucine encephalin (2 μg/mL)	15	

Table 3
Representative TWIMS settings for SYNAPT HDMS systems

	Triwave DC										
	Trap DC				IMS DC					Transfer DC	
Polarity	Entrance	Bias	Trap DC	Exit	Entrance	Helium cell DC	Helium exit	Bias	Exit	Entrance	Exit
ES+	3	45	0	0	25	35	−5	3	0	4	15
ES−	3	45	0	0	25	35	−5	3	0	4	15

	Gas controls		Triwave					
			Trap		IMS		Transfer	
Polarity	IMS gas (nitrogen) (mL/min)	Helium cell (mL/min)	Wave velocity (m/s)	Wave height (V)	Wave velocity (m/s)	Wave height (V)	Wave velocity (m/s)	Wave height (V)
ES+	90	180	311	6	600	40	220	4
ES−	90	180	311	6	600	40	220	4

3.2 Acquiring Information for Collision Cross Sections

After ionization, lipid ions cross the IM separation cell before MS detection. On the basis of the characteristic time required for a lipid ion to cross the IM separation cell (i.e., the drift time), it is possible to calculate the rotationally averaged collision cross sections (CCS), a physicochemical property, which is related to the lipid's charge, size, and shape, as well as the nature of the gas used in the IM cell.

CCS values can be automatically calculated from drift-time data by means of informatics solutions [27, 28]. Lipids can be annotated for both accurate masses and CCS values, thus creating a searchable database for lipidomics applications to increase the confidence of lipid identification and reducing the number of false positives [29]. Indeed, CCS values are independent of chromatographic conditions and analytical matrix, which might cause significant variations in retention times [28].

1. Set the instrument in Mobility TOF and wait for at least 1 h before acquiring CCS information. This step will ensure to obtain a stable CCS calibration.

2. In MassLynx open Intellistart and select CCS calibration. Next, prepare polyAla at the concentration of 10 µg/mL (alternatively you can use CCSMajor Mix solution) and infuse the solution at a flow rate of 5 µL/min. Check that the beam is stable and that the all single charged polyAla oligomers [23, 28] are present. Then use Intellistart for acquiring calibration data automatically (*see* **Note 2**).

3. When acquiring data from real samples, CCS values will be automatically derived from the CCS calibration curve using informatics products such as Progenesis® QI or UNIFI (Waters Corporation) [27, 28]. The difference between reference CCS and experimental CCS values (ΔCCS) can be used as a contribution to the identification score for lipids.

3.3 Lipidomics by LC-IM-MS

Historically, chromatographic separation has been used to maximize lipid separation prior to MS detection. For example, reversed-phase UPLC has been extensively used to separate lipid species, mostly according to their hydrophobicity [30–33]. Yet even the most advanced chromatographic technique cannot completely separate the wide array of lipids in biological samples, an insufficiency that is exacerbated when short chromatographic runs are needed to increase throughput [34]. Because chromatographic separations occur in seconds and IM separations in milliseconds, IM can be efficiently coupled to chromatography, providing an additional degree of separation and increasing peak capacity as well as the specificity of lipid identification (Fig. 1) [8–10, 34–37].

3.3.1 UPLC Preparation

1. Prepare mobile phases A and B.

    (a) Mobile phase A: 10 mM ammonium formate with 0.1% formic acid in 40:60 (v/v) acetonitrile/water.

    (b) Mobile phase B: 10 mM ammonium formate with 0.1% formic acid in 10:90 (v/v) acetonitrile/isopropanol.

2. Insert the column in the column compartment and set the column temperature at 55 °C, the flow rate at 0.4 mL/min, the injection volume at 5 µL, and the autosampler temperature at 10 °C.

3. Set the gradient condition. Gradient: Initial conditions started with 40% B and immediately a linear gradient from 40% to 43% B in 2 min. In the following 0.1 min, the percentage of mobile phase B was increased to 50%. Over the next 9.9 min, the gradient was further ramped to 54% B, and the amount of mobile phase B was increased to 70% in 0.1 min. In the final part of the gradient, the % B was increased to 99% in 5.9 min. The eluent composition returned to the initial conditions in 0.1 min, and the column was equilibrated, under the initial conditions, for 1.9 min before the next injection. The total run time was 20.0 min.

4. Equilibrate the column running the gradient four times without injecting any sample.

3.3.2 Data-Independent Acquisition Coupled with IM

The use of IM-MS improves the data-independent acquisition (MS[E]) process for identifying lipids in complex mixtures. Indeed, the incorporation of IM in the MS[E] process allows a new acquisition mode, which is named HDMS[E]. In HDMS[E], co-eluting lipid precursor ions can be separated by ion-mobility before fragmentation, resulting in cleaner MS/MS product-ion spectra (Fig. 2) [8, 9, 12, 38, 39]. Furthermore, during a HDMS[E] experiment the precursor ions and the product ions are drift-time aligned, this provides additional information for data deconvolution.

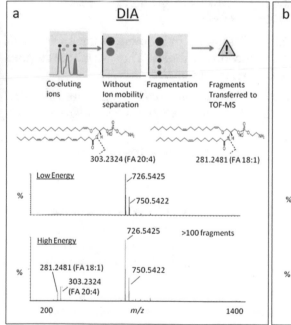
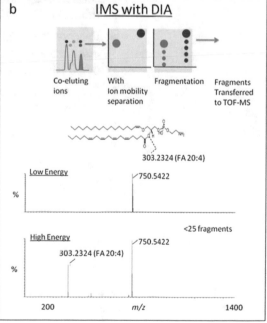

Fig. 2 Schematic visualization of acquisition using data-independent acquisition (MSE) and MSE coupled with IM (HDMSE). Combined with IM separation, fragmentation offers unique capabilities to increase specificity and confidence in identifying complex lipid structures Reproduced with permission from [25], © 2015, Springer

These cleaner MS/MS spectra facilitated lipid identification by increasing specificity and reducing false-positive assignments [8, 9, 12, 38, 39] (Fig. 2) (*see* **Note 3**).

1. Perform instrument performance checks, according to the manufacturer's guidelines.

2. Select the HDMS[E] and set the parameters of the two distinct functions (Fig. 2).

3. Function 1 uses low collision energy and therefore acquires information relative to the precursor ions. Use the parameter reported in Tables 2 and 3. We suggest setting collision energy in the trap cell and the transfer cell at 5 eV.

4. Function 2 uses high-collision energy and therefore acquires information relative to the fragment ions. Use the parameter reported in Tables 2 and 3. We suggest setting collision energy in the trap cell at 5 eV and ramping the collision energy from 20 to 40 eV in the transfer cell

3.3.3 Sample Analysis

1. Perform a lipid extraction procedure. Different procedures are available in the literature. They are based on type of sample (tissue, cells, plasma, etc.) or selected lipids of interest [40].

2. Prepare CCS QC standard solution by diluting LCMS QCRM mix 1:100 using 10% acetonitrile. Ensure precise CCS measurements by running the QC prior to sample analysis. Alternatively, the LipidoMix from Avanti can be used to determine the IM system suitability.

3. Prepare "pooled QC" samples by pooling a small aliquot of lipid extracts from individual samples. Thus, combine aliquots of 10 µL from each extracted sample.

4. Prepare a sample list. Include samples in a randomized order, a "pooled QC sample" and a CCS QC standard solution (*see* **Note 4**).

5. Inject pooled QC samples ten times, to stabilize the UPLC-IM-MS system.

6. Run the samples.

7. Use the pooled QC samples and CCS QC standard solution to verify the reproducibility of the analysis, and thus, evaluate retention time, intensity, and mass accuracy, and confirm that CCS (N_2) measurements fall within 2% of expected values.

3.3.4   Data Processing and Analysis

In a typical UPLC-IM-MS lipidomic experiment, a list of lipid annotations is automatically generated. The list includes, as molecular identifiers, CCS values as well as retention-time and mass (m/z) values [8, 23, 28, 41, 42]. Commercially available software, including Progenesis QI or UNIFI software, can be used to process and analyze TWIM-MS information, automatically deriving CCS values from the calibration curve.

Following is a typical procedure for searching CCS information:

1. Search lipids by matching experimentally derived CCS values against CCS databases [28]. Databases containing CCS for several lipid species are available in the literature [23, 28].

2. The difference between reference CCS (contained in the database) and experimental CCS values (ΔCCS) contributes to the identification score, in addition to accurate masses, isotopic pattern, and retention times. Fragmentation spectra can be analyzed in HDMS^E mode.

3. Lipids can be searched by selecting a ΔCCS less than that of a given threshold as tolerance parameters [23, 27, 28].

4. Filter and score identifications when querying the database with CCS information to reduce the number of false positives and negatives [28].

3.4 Lipidomics Imaging by IM-MS

MS imaging provides detailed spatial distribution of lipid species in biological tissues. Lipids are prominent features in MS imaging spectra of biological tissues under most desorption ionization sources, including matrix-assisted laser desorption ionization (MALDI) [9, 43] and desorption electrospray ionization (DESI) [44–46].

Lipids are localized across the surface of a sample in different compositions and concentrations. MS imaging makes possible topographic mapping of the lipid content of tissue sections without performing any type of extensive sample preparation, such as lipid extraction. In a typical MS imaging experiment, a focused excitatory beam (e.g., a laser or charged-solvent droplets) scans along all spatial axes of a tissue section [9, 47, 48]. Upon impact, lipid ions are desorbed and ionized from the sample surface and then sampled by a mass spectrometer.

The addition of IM to traditional MS imaging experiments allows separation of the lipid ions of interest from the interfering background before MS detection, resulting in a relatively improved signal-to-noise ratio and lipid localization (Fig. 3). In addition to accurate mass value, the characteristic CCS value associated with each desorbed lipid ion can be searched against databases, to support lipid identification [9]. Various desorption ionization techniques have been combined with M-MS for lipids imaging, among them matrix-assisted laser desorption ionization (MALDI) [9, 49] desorption electrospray ionization (DESI) (Figs. 1 and 3) and laser ablation electrospray ionization (LAESI) [50] (*see* **Note 5**).

3.4.1 Tissue Preparation for Lipidomics Imaging

Storage, chemical processing, and quality of tissue sectioning affect the quality of MS imaging results. The following tissue-processing procedure is recommended for obtaining optimal sections for DESI imaging:

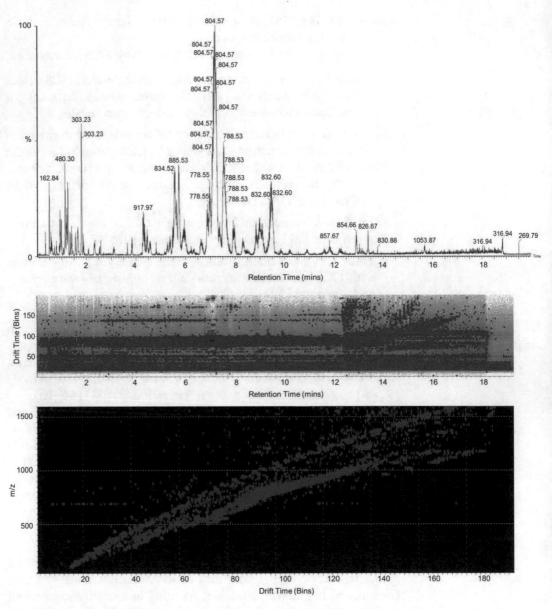

Fig. 3 Brain Porcine Lipid Extract analysis performed by UPLC-HDMS^E platform in positive mode. The first panel is the mass chromatogram, the second panel represents a 2D plot retention time vs drift time, while the third panel is 2D plot drift time vs m/z

1. For tissue sections, use fresh or flash frozen tissue stored at −80 °C (*see* **Note 6**).

2. Prepare 2.5% CMC solution in deionized water.

3. Add a few drops of CMC at the bottom of the cryomold (*see* **Note 7**).

4. Mount the specimen on the drop of CMC in the desired orientation (e.g., coronal or sagittal) (*see* **Note 8**).

5. Completely cover the tissue with CMC and snap freeze it.

6. Remove the CMC block and attach the *block to* the sample *mount of* a cryostat (e.g., CM3050 S, Leica Buffalo Grove, IL) using a couple of drops of CMC.

7. Typical for DESI imaging tissues are sectioned, between 10 and 20 μm in thickness. Sectioning can be done at approximately −22 °C.

8. Store tissue sections in an airtight, labeled slide holder placed inside a −80 °C freezer until further use.

9. Remove frozen tissue sections, transfer them to a vacuum desiccator, and allow about half hour to defrost them. Avoid condensation (*see* **Note 9**).

3.4.2 DESI Optimization

DESI-MS has been employed to detect lipid species from biological samples, including biofluids, mammalian tissue sections, microbial colonies, and cell cultures [51–54]. In DESI, lipids are desorbed and ionized from a sample surface by pneumatically assisted electrospray at atmospheric pressure or ambient conditions. To obtain DESI images, mass spectra are collected in a series of coordinates or pixels, which are later rendered into false-color images according to the intensity of ions. A motorized stage with a sample is moved, in two dimensions, to obtain mass spectrum at each pixel.

A schematic of a DESI ion source appears in Fig. 4. Briefly, it depicts the side-view of an electrospray emitter with pneumatic gas flow assembly on a three-dimensional, translational stage. The DESI ion source is mounted atop a two-dimensional, translational stage in front of an ion transfer capillary of a mass spectrometer. The DESI emitter and MS-inlet capillary are kept stationary. The spatial resolution of the DESI image depends on electrospray impact and size and the speed of acquisition (step velocity).

Optimizing the DESI source is critical for obtaining a robust and stable signal. Ideally, signal optimization should be performed using reference samples as close as possible to samples of interest. For instance, for tissue sections, optimize signal on adjacent serial sections.

1. Tune the mass spectrometer according to its manufacturer's instructions.

2. Optimize the DESI ion source (Prosolia) parameters listed in Table 4 for each set of samples by monitoring their signal intensity. Initially, optimize the MS signal by analyzing a spot drawn using a fine, red marker containing the rhodamine 6G dye (m/z 443) on a glass microscope slide.

3. Electrospray composition for DESI imaging depends on the analyte of interest and the tissue type. A typical composition of 98% methanol with 0.1% acetic acid *(v/v)* is recommended

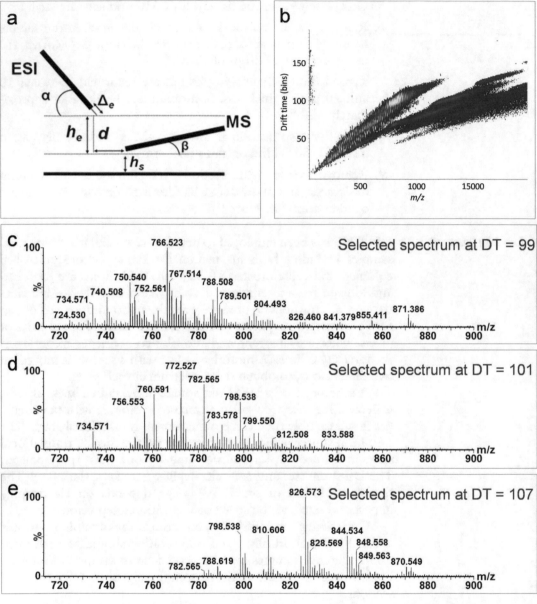

Fig. 4 Shown are (**a**) schematic of DESI ion source. Briefly, incidence angle of electrospray emitter (α), angle of ion collection capillary (β), height difference between emitter and ion collection capillary (h_e), distance between the emitter and MS inlet capillary (*d*), distance between ion-collection capillary and sample surface (h_s). The electrospray emitter offset is given by (Δe). (**b**) Representative two-dimensional plot (drift time vs *m/z* from a positive-ion DESI-TWIMS-MS analysis of a portion of a coronal section of mouse brain; (**c** and **d**) representative selection of lipid spectra isolated by specific drift times: 99, 101, and 107 bins

as a starting point for DESI-imaging method development (*see* **Note 10**).

4. Set and adjust the nebulizing gas pressures, solvent flow rate, and applied voltage. Recommended initial settings for optimization appear in Table 4.

Table 4
Recommended DESI imaging settings

Settings	Positive ion mode	Negative ion mode
Electrospray voltage	5000 V	4250 V
Solvent flow rate	2 μL per minute	
Nebulizing gas pressure	4.5 bar	
MS scan rate	1 scan per second	
Electrospray incidence angle (α)	75°	
Ion collection capillary angle (β)	5°	
Emitter tip to sample surface distance (d)	2 mm	
Emitter tip to ion collection capillary distance (he)	6 mm	
Ion transfer capillary orifice to surface (hs)	~1 mm	
Offset of electrospray emitter (Δe)	0.5 mm	

5. Geometrically align the DESI electrospray ion source, as shown in Fig. 4. Adjust the electrospray incidence angle (α), angle of ion collecting inlet capillary (β), distance between emitter and sample surface (he+ hs), distance between the electrospray emitter and MS inlet capillary (d), distance between ion collection capillary and sample surface (hs). Start with recommended values provided in Table 4 (*see* **Note 11**).

3.4.3 DESI Imaging with IM

1. CCS calibration can be performed using DESI or, separately, using electrospray ionization as previously described.

2. Before imaging, scan an optical image of the specimen on the slide using a regular flatbed scanner. The optical image is used to define an area of DESI-MS imaging.

3. The translation stage used in imaging is controlled by the manufacturer's software, such as HDI Imaging software (Waters). Stages are programmed to raster in successive rows creating a two-dimensional image (*see* **Note 12**).

4. Select the area of imaging, define pixel sizes (equal value for x and y is recommended), input desired stage scan speed (e.g., 100 μm per second) and line spacing (e.g., 100 μm) between rows (*see* **Note 13**).

5. To obtain data in IM mode, ensure that the scan speed is not set lower than 0.3 spectrum per second; the recommended setting is 1 scan per second.

6. Acquire IM-MS imaging data. After the DESI-imaging data acquisition concludes, put the mass the spectrometer in Standby mode, stop the nitrogen gas flow, and turn off syringe pump.

7. Process mass-spectral data. Ions can be rendered on the basis of m/z ratio and drift times. Other analyses include hierarchal clustering of ions, which groups related ions, and spatial correlation analysis, which shows all other ions with similar or opposite spatial distribution.

8. Extract lipids present in selected drift-time regions using Driftscope software.

4   Notes

1. As an alternative, normal-phase chromatography, hydrophilic interaction chromatography (HILIC), and supercritical fluid chromatography (SFC) can be used to separate lipid classes according to their polarity [34, 55–59].

2. When poly-DL-alanine is used as the calibrant species in both ES+ and ES−, calibration can be performed using oligomers, from $n = 3$ to $n = 11$, covering a mass range from 231 Da to 799 Da and a CCS mobility range from 151 Å2 to 306 Å2 in ES+ and from 150 Å2 to 308 Å2 in ES−.

3. IM-MS experiments might add selectivity to the separation and detection of structural isomers, isobars, and conformers [60–64]. Derivatization methods and alternative ion-mobility gases have both been used to maximize the separation of isobaric and isomeric lipid species by IM-MS [14]. The derivatization increases the CCS of the isomers, affecting the interactions of the lipid ion with the drift gas and thus improving their separation in the ion-mobility cell. In addition, IM gases of different polarizabilities, as well as changes in the pressure of these gases, have also been used to separate lipid isomers [14, 65, 66].

4. If a manual CCS calibration is required, you might include three injection of poly-DL-alanine or CCS Major mix calibration solution for CCS calculations.

5. Real-Time IM-MS. For a high-throughput fingerprinting of biological samples, various ambient ionization MS (AIMS) techniques allow the rapid sampling and ionization of lipids directly from solid or liquid samples, without prior sample treatment [67–69].

6. If the tissue is frozen, allow it to reach the temperature (e.g., approximately −20 °C) of cryomicrotome before beginning sectioning.

7. The size cryomold and volume of CMC depends on the size of the tissue.

8. Label the orientation of the tissue on the mold before starting.

9. Leaving fresh tissue sections in a desiccator too long will cause flaking, because of excessive drying, and can result in analyte delocalization.

10. Load a generous volume of electrospray solvent, to last the total time of the DESI imaging run.

11. The instrumental interlock should automatically turn off the high voltage while accessing the micromanipulators during DESI source adjustments. Yet, it is a good and recommended practice to put the mass spectrometer in its Standby mode.

12. Refer to the software instruction for procedure for setting up a DESI imaging experiment.

13. Scan speed and spacing between lines (i.e., total number of lines) affects the total time of MS imaging. The true pixel size is slightly larger than the affected area, because of the spreading of spray.

Acknowledgments

We would like to thank Drs. David Grant, Andrea Armirotti, Will Thompson, Michal Kliman, Hans Vissers, Kevin Giles, Jonathan Williams, Nick Tomczyk, and Suraj Dhungana for discussions we found most enlightening.

References

1. Fahy E, Subramaniam S, Murphy RC, Nishijima M, Raetz CR, Shimizu T, Spener F, van Meer G, Wakelam MJ, Dennis EA (2009) Update of the LIPID MAPS comprehensive classification system for lipids. J Lipid Res 50(Suppl)S9–14. doi:10.1194/jlr.R800095-JLR200

2. Quehenberger O, Armando AM, Brown AH, Milne SB, Myers DS, Merrill AH, Bandyopadhyay S, Jones KN, Kelly S, Shaner RL, Sullards CM, Wang E, Murphy RC, Barkley RM, Leiker TJ, Raetz CR, Guan Z, Laird GM, Six DA, Russell DW, McDonald JG, Subramaniam S, Fahy E, Dennis EA (2010) Lipidomics reveals a remarkable diversity of lipids in human plasma. J Lipid Res 51(11)3299–3305. doi:10.1194/jlr.M009449

3. Quehenberger O, Dennis EA (2011) The human plasma lipidome. N Engl J Med 365(19)1812–1823. doi:10.1056/NEJMra1104901

4. Wenk MR (2005) The emerging field of lipidomics. Nat Rev Drug Discov 4(7)594–610. doi:10.1038/nrd1776

5. Witting M, Maier TV, Garvis S, Schmitt-Kopplin P (2014) Optimizing a ultrahigh pressure liquid chromatography-time of flight-mass spectrometry approach using a novel sub-2mum core-shell particle for in depth lipidomic profiling of Caenorhabditis elegans. J Chromatogr A 1359:91–99. doi:10.1016/j.chroma.2014.07.021

6. Brown HA, Murphy RC (2009) Working towards an exegesis for lipids in biology. Nat Chem Biol 5(9)602–606

7. Shevchenko A, Simons K (2010) Lipidomics: coming to grips with lipid diversity. Nat Rev Mol Cell Biol 11(8)593–598. doi:10.1038/nrm2934

8. Shah V, Castro-Perez JM, McLaren DG, Herath KB, Previs SF, Roddy TP (2013) Enhanced data-independent analysis of lipids using ion mobility-TOFMS(E) to unravel quantitative and qualitative information in human plasma. Rapid Commun Mass Spectrom: RCM 27(19)2195–2200. doi:10.1002/rcm.6675

9. Hart PJ, Francese S, Claude E, Woodroofe MN, Clench MR (2011) MALDI-MS imaging of lipids in ex vivo human skin. Anal Bioanal

Chem 401(1)115–125. doi:10.1007/s00216-011-5090-4

10. Damen CW, Isaac G, Langridge J, Hankemeier T, Vreeken RJ (2014) Enhanced lipid isomer separation in human plasma using reversed-phase UPLC with ion-mobility/high-resolution MS detection. J Lipid Res 55(8)1772–1783. doi:10.1194/jlr.D047795

11. Kim HI, Kim H, Pang ES, Ryu EK, Beegle LW, Loo JA, Goddard WA, Kanik I (2009) Structural characterization of unsaturated phosphatidylcholines using traveling wave ion mobility spectrometry. Anal Chem 81(20)8289–8297. doi:10.1021/ac900672a

12. Castro-Perez J, Roddy TP, Nibbering NM, Shah V, McLaren DG, Previs S, Attygalle AB, Herath K, Chen Z, Wang SP, Mitnaul L, Hubbard BK, Vreeken RJ, Johns DG, Hankemeier T (2011) Localization of fatty acyl and double bond positions in phosphatidylcholines using a dual stage CID fragmentation coupled with ion mobility mass spectrometry. J Am Soc Mass Spectrom 22(9)1552–1567. doi:10.1007/s13361-011-0172-2

13. Kaur-Atwal G, Reynolds JC, Mussell C, Champarnaud E, Knapman TW, Ashcroft AE, O'Connor G, Christie SD, Creaser CS (2011) Determination of testosterone and epitestosterone glucuronides in urine by ultra performance liquid chromatography-ion mobility-mass spectrometry. Analyst 136(19)3911–3916. doi:10.1039/c1an15450h

14. Ahonen L, Fasciotti M, Gennas GB, Kotiaho T, Daroda RJ, Eberlin M, Kostiainen R (2013) Separation of steroid isomers by ion mobility mass spectrometry. J Chromatogr A 1310:133–137. doi:10.1016/j.chroma.2013.08.056

15. Dong L, Shion H, Davis RG, Terry-Penak B, Castro-Perez J, van Breemen RB (2010) Collision cross-section determination and tandem mass spectrometric analysis of isomeric carotenoids using electrospray ion mobility time-of-flight mass spectrometry. Anal Chem. doi:10.1021/ac101974g

16. Domalain V, Hubert-Roux M, Lange CM, Baudoux J, Rouden J, Afonso C (2014) Use of transition metals to improve the diastereomers differentiation by ion mobility and mass spectrometry. J Mass Spectrom: JMS 49(5)423–427. doi:10.1002/jms.3349

17. Kanu AB, Dwivedi P, Tam M, Matz L, Hill HH Jr (2008) Ion mobility-mass spectrometry. J Mass Spectrom: JMS 43(1)1–22. doi:10.1002/jms.1383

18. Lapthorn C, Pullen F, Chowdhry BZ (2013) Ion mobility spectrometry-mass spectrometry (IMS-MS) of small molecules: separating and assigning structures to ions. Mass Spectrom Rev 32(1)43–71. doi:10.1002/mas.21349

19. Fenn LS, Kliman M, Mahsut A, Zhao SR, McLean JA (2009) Characterizing ion mobility-mass spectrometry conformation space for the analysis of complex biological samples. Anal Bioanal Chem 394(1)235–244. doi:10.1007/s00216-009-2666-3

20. Kliman M, May JC, McLean JA (2011) Lipid analysis and lipidomics by structurally selective ion mobility-mass spectrometry. Biochim Biophys Acta 1811(11)935–945. doi:10.1016/j.bbalip.2011.05.016

21. Woods AS, Ugarov M, Egan T, Koomen J, Gillig KJ, Fuhrer K, Gonin M, Schultz JA (2004) Lipid/peptide/nucleotide separation with MALDI-ion mobility-TOF MS. Anal Chem 76(8)2187–2195. doi:10.1021/ac035376k

22. Dwivedi P, Schultz AJ, Hill HH (2010) Metabolic profiling of human blood by high resolution Ion Mobility Mass Spectrometry (IM-MS) Int J Mass Spectrom 298(1-3)78–90. doi:10.1016/j.ijms.2010.02.007

23. Paglia G, Angel P, Williams JP, Richardson K, Olivos HJ, Thompson JW, Menikarchchi L, Lai S, Walsh C, Moseley A, Plumb RS, Grant DF, Palsson BO, Langridge J, Geromanos S, Astarita G (2015) Ion-mobility-derived collision cross section as an additional measure for lipid fingerprinting and identification. Anal Chem 87(2)1137–1144. doi:10.1021/ac503715v

24. Jackson SN, Ugarov M, Post JD, Egan T, Langlais D, Schultz JA, Woods AS (2008) A study of phospholipids by ion mobility TOFMS. J Am Soc Mass Spectrom 19(11)1655–1662. doi:10.1016/j.jasms.2008.07.005

25. Shvartsburg AA, Isaac G, Leveque N, Smith RD, Metz TO (2011) Separation and classification of lipids using differential ion mobility spectrometry. J Am Soc Mass Spectrom 22(7)1146–1155. doi:10.1007/s13361-011-0114-z

26. Giles K, Pringle SD, Worthington KR, Little D, Wildgoose JL, Bateman RH (2004) Applications of a travelling wave-based radio-frequency-only stacked ring ion guide. Rapid Commun Mass Spectrom: RCM 18(20)2401–2414. doi:10.1002/rcm.1641

27. Paglia G, Menikarchchi L, Langridge J, Astarita G (2014) Travelling-Wave Ion Mobility-MS in metabolomics: workflows and bioinformatic tools. In: Rudaz S (ed) Identification and data processing methods in metabolomics. Future Medicine, London, UK. doi:10.4155/FSEB2013.14.224

28. Paglia G, Williams JP, Menikarchchi L, Thompson JW, Tyldesley-Worster R, Halldorsson S, Rolfsson O, Moseley A, Grant D, Langridge J, Palsson BO, Astarita G (2014) Ion mobility derived collision cross sections to

support metabolomics applications. Anal Chem 86(8)3985–3993. doi:10.1021/ac500405x

29. Paglia G, Williams JP, Menikarachchi L, Thompson JW, Tyldesley-Worster R, Halldórsson S, Rolfsson O, Moseley A, Grant D, Langridge J, Palsson BO, Astarita G (2014) Ion Mobility-derived Collision Cross-Sections to Support Metabolomics Applications. Analytical chemistry Accepted

30. Damen CW, Isaac G, Langridge J, Hankemeier T, Vreeken RJ (2014) Enhanced lipid isomer separation in human plasma using reversed-phase UPLC with ion-mobility/high-resolution MS detection. J Lipid Res. doi:10.1194/jlr.D047795

31. Churchwell MI, Twaddle NC, Meeker LR, Doerge DR (2005) Improving LC-MS sensitivity through increases in chromatographic performance: comparisons of UPLC-ES/MS/MS to HPLC-ES/MS/MS. J Chromatogr B Analyt Technol Biomed Life Sci 825(2)134–143. doi:10.1016/j.jchromb.2005.05.037

32. Plumb RS, Johnson KA, Rainville P, Smith BW, Wilson ID, Castro-Perez JM, Nicholson JK (2006) UPLC/MS(E); a new approach for generating molecular fragment information for biomarker structure elucidation. Rapid Commun Mass Spectrometry: RCM 20(13)1989–1994. doi:10.1002/rcm.2550

33. Swartz ME (2005) UPLC™: an introduction and review. J Liq Chromatogr Relat Technol 28:1253–1263

34. Cajka T, Fiehn O (2014) Comprehensive analysis of lipids in biological systems by liquid chromatography-mass spectrometry. Trends Anal Chem: TRAC 61:192–206. doi:10.1016/j.trac.2014.04.017

35. Campuzano I, Bush MF, Robinson CV, Beaumont C, Richardson K, Kim H, Kim HI (2012) Structural characterization of drug-like compounds by ion mobility mass spectrometry: comparison of theoretical and experimentally derived nitrogen collision cross sections. Anal Chem 84(2)1026–1033. doi:10.1021/ac202625t

36. Dear GJ, Munoz-Muriedas J, Beaumont C, Roberts A, Kirk J, Williams JP, Campuzano I (2010) Sites of metabolic substitution: investigating metabolite structures utilising ion mobility and molecular modelling. Rapid Commun Mass Spectrom: RCM 24(21)3157–3162. doi:10.1002/rcm.4742

37. Wickramasekara SI, Zandkarimi F, Morre J, Kirkwood J, Legette L, Jiang Y, Gombart AF, Stevens JF, Maier CS (2013) Electrospray quadrupole travelling wave ion mobility time-of-flight mass spectrometry for the detection of plasma metabolome changes caused by xantho-humol in obese zucker (fa/fa) rats. Metabolites 3(3)701–717. doi:10.3390/metabo3030701

38. Chong WP, Goh LT, Reddy SG, Yusufi FN, Lee DY, Wong NS, Heng CK, Yap MG, Ho YS (2009) Metabolomics profiling of extracellular metabolites in recombinant Chinese Hamster Ovary fed-batch culture. Rapid Commun Mass Spectrom: RCM 23(23)3763–3771. doi:10.1002/rcm.4328

39. Gonzales GB, Raes K, Coelus S, Struijs K, Smagghe G, Van Camp J (2014) Ultra(high)-pressure liquid chromatography-electrospray ionization-time-of-flight-ion mobility high definition mass spectrometry for the rapid identification and structural characterization of flavonoid glycosides from cauliflower waste. J Chromatogr A 1323:39–48. doi:10.1016/j.chroma.2013.10.077

40. Armstrong D (2010) Lipidomics : volume 2: methods and protocols, Methods in Molecular Biology, Spinger

41. Malkar AD, Devenport NA, Martin HJ, Patel P, Turner MA, Watson P, Maughan RJ, Reid HJ, Sharp BL, CLP T, Reynolds JC, Creaser CS (2013) Metabolic profiling of human saliva before and after induced physiological stress by ultra-high performance liquid chromatography–ion mobility–mass spectrometry. Metabolomics 9(6)1192–1201. doi:10.1007/s11306-013-0541-x

42. Harry EL, Weston DJ, Bristow AW, Wilson ID, Creaser CS (2008) An approach to enhancing coverage of the urinary metabonome using liquid chromatography-ion mobility-mass spectrometry. J Chromatogr B Analyt Technol Biomed Life Sci 871(2)357–361. doi:10.1016/j.jchromb.2008.04.043

43. Fang L, Harkewicz R, Hartvigsen K, Wiesner P, Choi SH, Almazan F, Pattison J, Deer E, Sayaphupha T, Dennis EA, Witztum JL, Tsimikas S, Miller YI (2010) Oxidized cholesteryl esters and phospholipids in zebrafish larvae fed a high cholesterol diet: macrophage binding and activation. J Biol Chem 285(42)32343–32351. doi:10.1074/jbc.M110.137257

44. Balas L, Guichardant M, Durand T, Lagarde M (2014) Confusion between protectin D1 (PD1) and its isomer protectin DX (PDX). An overview on the dihydroxy-docosatrienes described to date. Biochimie 99:1–7. doi:10.1016/j.biochi.2013.11.006

45. Chen P, Fenet B, Michaud S, Tomczyk N, Vericel E, Lagarde M, Guichardant M (2009) Full characterization of PDX, a neuroprotectin/protectin D1 isomer, which inhibits blood platelet aggregation. FEBS Lett 583(21)3478–3484. doi:10.1016/j.febslet.2009.10.004

46. Skraskova K, Claude E, Jones EA, Towers M, Ellis SR, Heeren RM (2016) Enhanced capabilities for imaging gangliosides in murine brain with matrix-assisted laser desorption/ionization and desorption electrospray ionization mass spectrometry coupled to ion mobility separation. Methods. doi:10.1016/j.ymeth.2016.02.014

47. Trim PJ, Atkinson SJ, Princivalle AP, Marshall PS, West A, Clench MR (2008) Matrix-assisted laser desorption/ionisation mass spectrometry imaging of lipids in rat brain tissue with integrated unsupervised and supervised multivariant statistical analysis. Rapid Commun Mass Spectrom: RCM 22(10)1503–1509. doi:10.1002/rcm.3498

48. Djidja MC, Claude E, Snel MF, Francese S, Scriven P, Carolan V, Clench MR (2010) Novel molecular tumour classification using MALDI-mass spectrometry imaging of tissue microarray. Anal Bioanal Chem 397(2)587–601. doi:10.1007/s00216-010-3554-6

49. Galhena AS, Harris GA, Kwasnik M, Fernandez FM (2010) Enhanced direct ambient analysis by differential mobility-filtered desorption electrospray ionization-mass spectrometry. Anal Chem 82(22)9159–9163. doi:10.1021/ac102340h

50. Li H, Smith BK, Márk L, Nemesa P, Nazarian J, Vertes A (2014) Ambient molecular imaging by laser ablation electrospray ionization mass spectrometry with ion mobility separation. Int J Mass Spectrom 377:681–689

51. Takáts Z, Wiseman JM, Gologan B, Cooks RG (2004) Mass spectrometry sampling under ambient conditions with desorption electrospray ionization. Science 306(5695)471–473. doi:10.1126/science.1104404

52. Wiseman JM, Ifa DR, Song Q, Cooks RG (2006) Tissue imaging at atmospheric pressure using desorption electrospray ionization (DESI) mass spectrometry. Angew Chem Int Ed 45(43)7188–7192

53. Song Y, Talaty N, Tao WA, Pan Z, Cooks RG (2007) Rapid ambient mass spectrometric profiling of intact, untreated bacteria using desorption electrospray ionization. Chem Commun 1:61–63. doi:10.1039/B615724F

54. Watrous J, Hendricks N, Meehan M, Dorrestein PC (2010) Capturing bacterial metabolic exchange using thin film desorption electrospray ionization-imaging mass spectrometry. Anal Chem 82(5)1598–1600. doi:10.1021/ac9027388

55. Jones MD, Rainville PD, Isaac G, Wilson ID, Smith NW, Plumb RS (2014) Ultra high resolution SFC-MS as a high throughput platform for metabolic phenotyping: application to metabolic profiling of rat and dog bile. J Chromatogr B Analyt Technol Biomed Life Sci 966:200–207. doi:10.1016/j.jchromb.2014.04.017

56. Novakova L, Chocholous P, Solich P (2014) Ultra-fast separation of estrogen steroids using subcritical fluid chromatography on sub-2-micron particles. Talanta 121:178–186. doi:10.1016/j.talanta.2013.12.056

57. Lee JW, Nagai T, Gotoh N, Fukusaki E, Bamba T (2014) Profiling of regioisomeric triacylglycerols in edible oils by supercritical fluid chromatography/tandem mass spectrometry. J Chromatogr B Analyt Technol Biomed Life Sci 966:193–199. doi:10.1016/j.jchromb.2014.01.040

58. Lee JW, Nishiumi S, Yoshida M, Fukusaki E, Bamba T (2013) Simultaneous profiling of polar lipids by supercritical fluid chromatography/tandem mass spectrometry with methylation. J Chromatogr A 1279:98–107. doi:10.1016/j.chroma.2013.01.020

59. Bamba T, Lee JW, Matsubara A, Fukusaki E (2012) Metabolic profiling of lipids by supercritical fluid chromatography/mass spectrometry. J Chromatogr A 1250:212–219. doi:10.1016/j.chroma.2012.05.068

60. Clemmer DE, Jarrold MF (1997) Ion mobility measurements and their applications to clusters and biomolecules. J Mass Spectrom 32:577–592

61. Pringle SD, Giles K, Wildgoose JL, Williams JP, Slade SE, Thalassinos K, Bateman RH, Bowers MT, Scrivens JH (2007) An investigation of the mobility separation of some peptide and protein ions using a new hybrid quadrupole/travelling wave IMS/oa-ToF instrument. Int J Mass Spectrom 261(1)1–12

62. McLean JR, McLean JA, Wu Z, Becker C, Perez LM, Pace CN, Scholtz JM, Russell DH (2010) Factors that influence helical preferences for singly charged gas-phase peptide ions: the effects of multiple potential charge-carrying sites. J Phys Chem B 114(2)809–816. doi:10.1021/jp9105103

63. Bohrer BC, Merenbloom SI, Koeniger SL, Hilderbrand AE, Clemmer DE (2008) Biomolecule analysis by ion mobility spectrometry. Annu Rev Anal Chem 1:293–327. doi:10.1146/annurev.anchem.1.031207.113001

64. Shrestha B, Vertes A (2014) High-throughput cell and tissue analysis with enhanced molecular coverage by laser ablation electrospray ionization mass spectrometry using ion mobility separation. Anal Chem 86(9)4308–4315. doi:10.1021/ac500007t

65. Fasciotti M, Sanvido GB, Santos VG, Lalli PM, McCullagh M, de Sa GF, Daroda RJ, Peter MG, Eberlin MN (2012) Separation of isomeric disaccharides by traveling wave ion mobility mass spectrometry using CO2 as drift gas. J Mass Spectrom: JMS 47(12)1643–1647. doi:10.1002/jms.3089

66. Lalli PM, Corilo YE, Fasciotti M, Riccio MF, de Sa GF, Daroda RJ, Souza GH, McCullagh M, Bartberger MD, Eberlin MN, Campuzano ID (2013) Baseline resolution of isomers by traveling wave ion mobility mass spectrometry: investigating the effects of polarizable drift gases and ionic charge distribution. J Mass Spectrom: JMS 48(9)989–997. doi:10.1002/jms.3245

67. Wu C, Dill AL, Eberlin LS, Cooks RG, Ifa DR (2013) Mass spectrometry imaging under ambient conditions. Mass Spectrom Rev 32(3)218–243. doi:10.1002/mas.21360

68. Paglia G, D'Apolito O, Gelzo M, Dello Russo A, Corso G (2010) Direct analysis of sterols from dried plasma/blood spots by an atmospheric pressure thermal desorption chemical ionization mass spectrometry (APTDCI-MS) method for a rapid screening of Smith-Lemli-Opitz syndrome. Analyst 135(4)789–796

69. Paglia G, Ifa DR, Wu C, Corso G, Cooks RG (2010) Desorption electrospray ionization mass spectrometry analysis of lipids after two-dimensional high-performance thin-layer chromatography partial separation. Anal Chem 82(5)1744–1750. doi:10.1021/ac902325j

Chapter 6

In Vitro Assay to Extract Specific Lipid Types from Phospholipid Membranes Using Lipid-Transfer Proteins: A Lesson from the Ceramide Transport Protein CERT

Kentaro Hanada and Toshihiko Sugiki

Abstract

Several mechanisms deliver specific lipid types from one organelle to another in cells. Two distinct mechanisms exist for inter-organelle lipid trafficking in eukaryotic cells: vesicular and non-vesicular. Lipid-transfer proteins (LTPs) that catalyze the inter-membrane transfer of lipids play pivotal roles in non-vesicular systems. As a key biochemical feature, LTPs extract a specific lipid type from membranes in vitro. This chapter describes two assay systems to assess the lipid extraction activity of the ceramide transport protein CERT from membranes under cell-free conditions: one is a conventional assay system using radioactive lipid ligands, while the other is a more recently developed assay system using surface plasmon resonance. These methods are applicable to other LTPs.

Key words Lipid-transfer proteins, Lipid extraction, Membranes, Ceramide, CERT, START domain, Surface plasmon resonance

1 Introduction

Lipids are the major constituents of all cell membranes and play dynamic roles in organelle structures and functions. In eukaryotic cells, an organelle in which a lipid type is synthesized is often different from the site at which that lipid exerts its functions. Thus, cells utilize various systems to deliver specific lipid types to their appropriate destinations. Two distinct mechanisms exist for inter-organelle lipid trafficking in eukaryotic cells: vesicular and non-vesicular. Many lipid-transfer proteins (LTPs) have been shown to play pivotal roles in the non-vesicular mechanisms identified to date [1, 2].

As depicted in the nomenclature, LTPs catalyze the inter-membrane transfer of lipids in vitro, and, in many cases, LTP-mediated transfer may occur in the absence of biological energy such as ATP. Therefore, LTPs are capable of transferring lipids between

Paul Wood (ed.), *Lipidomics*, Neuromethods, vol. 125,
DOI 10.1007/978-1-4939-6946-3_6, © Springer Science+Business Media LLC 2017

different membranes until an equilibrated state is reached. However, the LTP-mediated inter-organelle transport of lipids appears to be coupled with the consumption of biological energy such as ATP in living cells [3, 4], thereby leading to unidirectional lipid transport. During the inter-membrane transfer of lipids, the energetically highest barrier process is the most likely to extract a hydrophobic molecule (lipid) from a hydrophobic environment (membrane) into an aqueous phase (the cytosol in cells) [5, 6]. Therefore, as a key biochemical feature, LTPs are capable of extracting a specific lipid type from membranes in vitro.

Not all biochemically annotated LTPs participate in the inter-membrane transport of lipids in cells. For example, the yeast Sec14 protein, which has been identified as a phospholipid-transfer protein, appears to use its lipid ligands as functional modulators, not transport substrates of the protein [7]. Furthermore, the physiological function of intra-lysosomal sphingolipid activator proteins, which mediate inter-membrane sphingolipid transfer in vitro, is to transfer specific sphingolipids from the lysosomal membrane to soluble hydrolases in order to facilitate the degradation of sphingolipids in the lumen of lysosomes [8]. Such "apparent LTPs" may also exhibit lipid-extracting activity in vitro. Hence, cell-free assay systems to assess the ability of proteins to extract specific lipid types from membranes with rational expandability may represent an important basic tool in the molecular biology of lipids.

We herein described two assay systems to assess the lipid extraction activity of the ceramide transport protein CERT, a typical LTP, from membranes under cell-free conditions: one is a conventional assay system using radioactive lipid ligands [9], while the other is a more recently developed assay system using surface plasmon resonance (SPR) [10]. Although the principle and procedure of the conventional system are simple, its expansion to various lipid types may be practically limited by the availability of radioactive lipids. However, this limitation was overcome by the recently developed system, in which the release of lipid molecules from a phospholipid membrane matrix is monitored as a loss in weight from the matrix set on an SPR sensor. In addition, the types of lipids extracted from the matrix are confirmed by a mass spectrometry (MS) analysis of the out-flow fluid of the SPR assay. Therefore, investigators may select one of the two systems or others based on the purpose of the experiments and also the availability of materials.

2 Materials

The water used is deionized water purified by the Milli-Q system (Merck Millipore). The chemical reagents used are of analytical grade.

2.1 Purification of Recombinant CERT and Its Derivatives from Bacterial Cultures

1. The *Escherichia coli* (*E. coli*) BL21 (DE3) strain for the expression of hexahistidine (*His*$_6$)-tagged human CERT or its derivatives: *E. coli* BL21 (DE3) cells are transfected with the pET-28a(+) plasmid (Novagen), in which the cDNA coding the CERT protein or its derivatives (CERTΔST is a START domain-deleted CERT while ST CERT is a construct of the START domain only) is inserted in-frame by a heat shock method, and transformed as a bacterium resistant to kanamycin (25 μg/mL).

2. Luria-Bertani (LB) broth containing kanamycin: A stock solution of kanamycin (50 mg/mL; sterilized by filtration) is added to pre-autoclaved LB broth (trypeptone 10 g, yeast extract 5 g, and NaCl 5 g in 1 L of water) at a final concentration of 25 μg/mL. Store at room temperature.

3. 1 M Isopropyle-1-thio-β- D -galactopyranoside (IPTG): The stock solution of 1 M IPTG is stored at −20 °C.

4. *E. coli* lysis buffer: *E. coli* lysis buffer consists of 25 mM Tris–HCl (pH 7.4), 1% Triton X-100, 1 mM orthovanadic acid, 50 mM sodium fluoride, 5 mM sodium pyrophosphoric acid, 2.5 mM 2-mercaptoethanol, 0.27 M sucrose, and protease inhibitors (one tablet of EDTA-free Complete™ protease inhibitor cocktail, Roche Diagnostics, per 50 mL) [*see* **Note 1**]. Prepare freshly before use or store at 20 °C and use within ~2 weeks.

5. Probe-type sonicator: Model W-225R, Heat Systems manufactured by Ultrasonics Inc. or its functional equivalent.

6. Ultracentrifuge, rotor, and tube: Centrifuge machine, Himac CS120EX (HITACHI Engineering Machine, Co., Ltd); centrifugation rotor, S100AT5; centrifugation tube, 4 mL polycarbonate tube (4PC); or their functional equivalents.

7. TALON® metal affinity resin: TALON® metal affinity resin, a cobalt ion chelate resin (Clontech, Co., Ltd) is used for the affinity chromatography of *His*$_6$-tagged CERT.

8. TALON®-equilibration buffer (TEB): TEB consists of 50 mM sodium phosphate buffer (pH 7.0) and 0.3 M NaCl [*see* **Note 2**]. Store at 4 °C (after autoclaving for long storage).

9. Imidazole stock solution: The stock solution of 1 M Imidazole is stored at 4 °C.

10. Dialysis bag and buffer: The cellulose ester membrane manufactured by Spectrum Laboratories Inc. is used as a dialysis bag. The membrane is prewashed as described in the manufacturer's manual. The prewashed membrane is stored in 0.1% (w/v) NaN$_3$ at 4 °C, and thoroughly rinsed with water before use. The dialysis buffer is 10 mM Tris–HCl (pH 7.4) containing 0.25 M sucrose, or 10 mM Tris–HCl (pH 7.4) containing 0.15 M NaCl.

11. Protein quantification: The bicinchoninic acid protein (BCA) protein assay kit (Thermo Scientific) or its equivalent is used for protein quantification.

2.2 Extraction of Lipids from Phospholipid Membranes by CERT: Conventional Assay Using Radioactive Lipid Ligands

1. Nonradioactive lipids: Egg yolk phosphatidylcholine (PC) and phosphatidylethanolamine (PE) are from Avanti-Polar Lipids. Synthesized phospholipids (e.g., *sn*-1-palmitoyl-2-oleoyl-PC and -PE) may be used as alternatives.

2. Radioactive lipids: [*Palmitoyl*-1-^{14}C] *N*-palmitoyl-D-*erythro*-sphingosine (55 mCi/mmol), [*oleoyl*-1-^{14}C]dioleoyl-*rac*-glycerol (55 mCi/mmol), [*cholinemethyl*-^{14}C]sphingomyelin (55 mCi/mmol), [*dipalmitoyl*-1-^{14}C] L-α-dipalmitoylphosphatidylcholine (55 mCi/mmol), and D-*erythro* -[3-^{3}H]sphingosine (20 Ci/mmol) are from American Radiolabeled Chemicals. [1α,2α(n)-^{3}H]cholesterol (49 Ci/mmol) is from Amersham Bioscience. These lipids are solved in volatile organic solvents.

3. Buffers.
   Buffer 1: 20 mM Hepes/NaOH buffer (pH 7.4) containing 50 mM NaCl and 1 mM EDTA. Store at 4 °C (or at −20 °C for long storage).
   Buffer 2: 50 mM Hepes/NaOH buffer (pH 7.4) containing 100 mM NaCl and 0.5 mM EDTA.

4. Microfuge tubes: Polypropylene 1.5 mL microfuge tubes with a safe-lock are from Eppendorf AG.

5. Bath-type sonicator: Model 2210 manufactured by Branson, Co., Ltd. or its functional equivalent.

6. Ultracentrifuge, rotor, and tube: We use the following set to precipitate phospholipid membranes by ultracentrifugation: centrifuge machine, Himac CS120EX (HITACHI Engineering Machine, Co., Ltd); centrifugation rotor, RP100AT3; centrifugation tube, 0.23 mL polycarbonate tube (0.23PC); or their functional equivalents.

7. Nitrogen gas: Nitrogen gas in a bomb with an adjuster.

8. Chemical hood: Organic solvents are handled in a chemical hood.

2.3 Extraction of Ceramide from Phospholipid Membranes by the START Domain of CERT: Assay Using SPR

1. Nonradioactive lipids: Purified 1-palmitoyl-2-oleoyl-phosphatidylcholine (POPC), 1-palmitoyl-2-oleoyl-phosphatidylethanolamine(POPE),and1-palmitoyl-2-oleoyl-phosphatidylserine (POPS) are purchased in powder form from Avanti-Polar Lipids, and *N*-palmitoylsphingosine (C$_{16}$-ceramide) is from Biomol International.

2. Organic solvent: Chloroform/methanol mixture (1:1 [v/v]). It can be stored at −30 °C in an explosion-proof freezer.

3. Pyrex® glass test tubes (16 × 100 mm, 11 mL) with polytetrafluoroethylene (Teflon™)-lined screw caps are from IWAKI Pyrex.

4. Solutions:

   (a) Milli-Q water.

   (b) 20 mM 3-[(3-cholamidopropyl)dimethylammonio]pro-panesulfonate (CHAPS): Store in a dark place at room temperature.

   (c) SPR running buffer: 10 mM Hepes/NaOH buffer (pH 7.3) containing 100 mM NaCl and 5 mM TCEP/HCl [see **Note 3**].

   (d) 20 mM sodium hydroxide (NaOH) is prepared by diluting 1 M NaOH stock solution with Milli-Q water. It can be stored at room temperature for a year.

   (e) 0.1 mg/mL bovine serum albumin (BSA) solution is prepared prior to starting the SPR experiments by diluting 2 mg/mL BSA stock solution (purchased from Nacalai Tesque, Inc.) with the SPR running buffer.

   The SPR running buffer and 0.1 mg/mL BSA solutions are prepared prior to the start of SPR experiments; however, the former can be stored at 4 °C for several months. All the solutions applied to the SPR machine are pre-filtrated by passing them through a filter membrane (pore size: 0.22 μm), and also degassed by sonicating the filtrated solutions in the containers placed in a bath-type sonicator under reduced pressure with suction.

5. Microfuge tubes: Polypropylene 1.5 mL microfuge tubes with a safe-lock are from Eppendorf AG.

6. Glass vials: Flat-bottom and clear glass liquid scintillation vials with screw caps [see **Note 4**].

7. Mini-Extruder: All components, which are polycarbonate membranes (pore size: 50 nm), filter-supporting rough membranes, syringe holding/heating block, Teflon™-based membrane supporter and outer cases to fix the complex of membrane/Teflon™ modules, two gastight syringes (1000 μL) along with Teflon™-tipped plungers, and others are available as a kit from Avanti-Polar Lipids.

8. Constant-temperature incubator: Panasonic MIR-162 or its functional equivalent is needed in order to maintain the temperature of the syringe holding/heating block of the Mini-Extruder at 37 °C during this process [see **Note 5**].

9. Dried nitrogen gas [see **Note 6**]: In order to moderate and finely adjust the output gas flow speed, other accessories such as a gas-pressure regulator, pressure-proof hard tube, and adequate needle (briefly, a disposable Pasteur glass pipette is acceptable) are also necessary.

10. Sample lyophilizer with vacuum/freeze drying: EYELA FD0830 and the external vacuum pump TAITEC GCD-136XA, or comparable products.

11. −80 °C deep-freezer: Panasonic MDF-DU300H or its comparable product.

12. SPR equipment: Experimental procedures using BIACORE 2000 or 3000 (GE Healthcare) are described in this study. High-end models such as Biacore T100 or S51 can be alternatives.

13. Sensor chip for SPR: L1 Sensor chips (GE Healthcare).

2.4 Confirmation of Lipid Types Extracted in the SPR Assay System by High-Performance Liquid Chromatography (HPLC) and MS

1. Standard lipid: C_{16}-ceramide is purchased in powder form from Biomol International.

2. Organic solvents:
   (a) Chloroform/methanol mixture (1:2 [v/v]). It can be stored at room temperature or at −30 °C in an explosion-proof freezer.
   (b) Hexane/ethanol/2-propanol (98.7:1:0.1 [v/v]). It is prepared prior to use. When used for HPLC, the organic solvent is pre-filtrated by passing it through a PVDF membrane (pore size: 0.22 μm) and also degassed by vacuuming it for ~10 min (long-term vacuuming is avoided in order to prevent over-volatilization of the organic solvent).

3. Pyrex® glass test tubes (16 × 100 mm, 11 mL) with Teflon™-lined screw caps are from IWAKI Pyrex.

4. HPLC system:
   (a) NanoSpace SI-2 (Shiseido Co., Ltd) or its comparable product.
   (b) Capcell Pak CN column (solid resin particles with a diameter of 5 μm, which have cyano groups, are filled in a 2.0 × 250 mm column) or its comparable cyano-bonded normal phase LC column.

5. MS system:
   (a) LCQ Advantage mass spectrometer (ThermoFisher Scientific Inc.) or its comparable electrospray-ionization (ESI)-MS equipment.
   (b) Evaporative light-scattering detector (ELSD) (Varian Inc.) or its comparable product.

3 Methods

3.1 Purification of Recombinant CERT

Recombinant *His₆*-tagged human CERT or its derivatives are expressed in *E. coli* and then purified by affinity chromatography. In order to prevent unwanted proteolysis and dephosphorylation, protease inhibitors and phosphatase inhibitors are included in buffers during the purification process. All manipulations are conducted on ice or 4 °C unless otherwise noted.

1. *E. coli* cells for the expression of *His₆*-tagged CERT or its derivatives are cultured at 37 °C in LB broth containing kanamycin until cell turbidity reaches 0.6 in absorbance at a wavelength of 600 nm.

2. When absorbance reaches 0.6, the culture is cooled to ~25 °C in a water bath, supplemented with IPTG at a final concentration of 250 µM, and further cultured at 25 °C overnight [*see* **Note 7**].

3. *E. coli* cells are harvested by centrifugation (1400 × g, 15 min).

4. After the removal of the supernatant, the precipitated cells are suspended in the *E. coli* lysis buffer.

5. The cell suspension (5 mL of which is dispensed into a 15 mL plastic tube) is frozen at −80 °C, thawed at room temperature, and sonicated eight times for 20 s at 1 min intervals with a probe-type sonicator at a 24 W output [*see* **Note 8**].

6. After high-speed centrifugation of the sonicated cell lysate (100,000 × g, 1 h), the supernatant fraction is collected.

7. During centrifugation of the cell lysate, TALON® resin (~2 mL of the bed volume for a typical experiment) is pre-equilibrated with TEB in a conical 50 mL tube. After brief centrifugation at a low speed (500 × g, 5 min), the supernatant buffer is removed. Pre-equilibrated TALON® resin is stored on ice or at 4 °C.

8. The collected supernatant fraction (~25 mL) is transferred to a 50 mL conical tube containing 2 mL of TALON® metal affinity resin pre-equilibrated with TEB.

9. After capping, the tube is repeatedly inverted with a rotary shaker for 1 h.

10. After brief centrifugation at a low speed (1000 × g, 3 min), the supernatant buffer is removed [*see* **Note 9**].

11. The precipitated TALON® resin is suspended with 20 mL of a wash buffer (TEB supplemented with 10 mM imidazole).

12. After low speed centrifugation (1000 × g, 3 min), the supernatant buffer is removed.

13. **Steps 10** and **11** are repeated two more times for washing.

14. The precipitated resin is transferred to a disposable column, and washed with 20 mL of the wash buffer.

15. The resin in the column is washed with 20 mL of the wash buffer.

16. Four milliliters of an elution buffer (TEB supplemented with 0.15 M imidazole) is added to the column, and the eluent is collected.

17. The eluent is dialyzed against 500 mL of dialysis buffer for ~1 h, and substituted with 1 L of the buffer overnight.

18. The dialyzed sample is collected, dispensed into 1.5 mL microfuge tubes, and stored at –80 °C. By using a small volume (10–20 μL) of the collected sample, its protein concentration is determined with a BCA protein assay kit using BSA as the standard.

3.2 CERT-Mediated Extraction of Various Lipid Types from Phospholipid Membranes: Conventional Assay Using Radioactive Lipid Ligands

This section describes an assay for the CERT-mediated extraction of radioactive ceramide from artificial phospholipid membrane vesicles. The principle of this assay is simple, as depicted in Fig. 1: When phospholipid membrane vesicles containing a trace amount of radioactive ceramide are centrifuged, ceramide is almost completely precipitated accompanied by membrane vesicles. However, when CERT is present, ceramide is released from the membrane vesicles and allocated to the supernatant fraction accompanying CERT, while phospholipid membrane vesicles are completely precipitated. The radioactivity of the supernatant fraction is a measure of the amount of ceramide extracted from the membranes by CERT. This assay is applicable to other lipid types if other radioactive lipid types are used instead of ceramide (Fig. 2) [*see* **Note 10**]. In order to estimate the background release of lipids, negative control experiments without proteins MUST be performed. In addition, we routinely conduct another control experiment with CERTΔST, in which the lipid-transfer START domain is deleted, to determine "lipid-transfer domain"-dependent activity (Fig. 2).

1. On the day of the assay, lipid membrane vesicles consisting of PC, PE, and [*palmitoyl*-1-^{14}C] *N*-palmitoyl-D-*erythro*-sphingosine

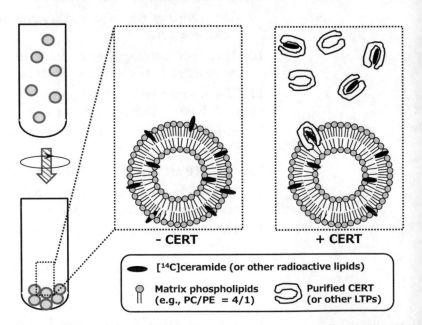

- CERT   + CERT

- [^{14}C]ceramide (or other radioactive lipids)
- Matrix phospholipids (e.g., PC/PE = 4/1)   Purified CERT (or other LTPs)

Fig. 1 Schematic diagram of the lipid extraction assay using radioactive ligand-embedded phospholipid vesicles

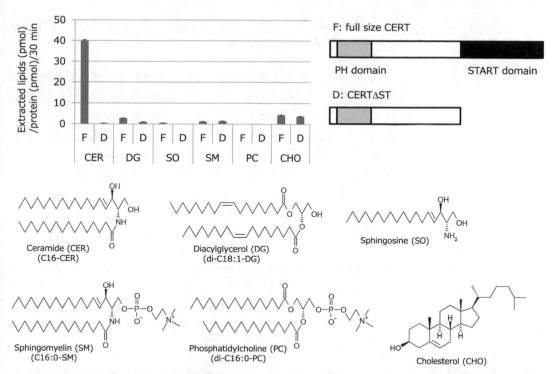

Fig. 2 Substrate specificity of lipid extraction from phospholipid vesicles by CERT. CERT extracts ceramide from the phospholipid membrane in a START domain-dependent manner. The domain structures of CERT and CERTΔST, and chemical structures of lipid ligands are shown. The data presented were published in Ref. 9

([^{14}C]ceramide) (800:200:3, [mol/mol]) are prepared by mild sonication as described below. Since 50 μg of lipid vesicles is used per assay under standard assay conditions, the investigator determines the amount of vesicles to be prepared.

2. Appropriate amounts of organic stock solutions of PC, PE, and [^{14}C]ceramide at a final molar ratio of 800:200:3 are placed in a 1.5 mL microfuge tube, and the mixture is dried by spraying nitrogen gas.

3. Buffer 1 is added to the dried lipid mixture so that the concentration becomes 2.5 mg /mL.

4. The supersonic treatment is gently performed using a bath-type sonicator. The supersonic treatment is performed at 25 °C, and the procedure involving the supersonic treatment for 3 min, the vortex for 30 s, and the supersonic treatment for 3 min is performed in this order. The lipid membrane prepared in this manner is maintained at 25 °C until used in the ceramide extraction reaction (**step 6**).

5. Purified CERT or its derivatives are used as the protein sample. Vehicle buffer is added without proteins for a protein-free background control. Under standard conditions, the amount of protein corresponding to 450 pmol (which is a twofold

molar equivalent amount of ceramide contained in the donating membrane) is adjusted to 30 μL using buffer 2 in a 0.23 mL ultracentrifugation tube.

6. The reaction is initiated by adding 20 μL of phospholipid membranes containing [^{14}C]ceramide [*see* **Note 11**].

7. This mixture is incubated at 37 °C for 30 min.

8. The mixture is centrifuged at 50,000 × *g* for 30 min at 4 °C and the phospholipid membrane is precipitated.

9. The supernatant is retrieved carefully by pipetting, and transferred into a scintillation counting vial.

10. After the addition of the scintillation cocktail (e.g., 2 mL of ACSII, GE), the radioactive activity of ^{14}C in the supernatant fraction is counted using a liquid scintillation counter.

11. Protein-dependent extraction activity is determined after subtracting the background control value.

12. The difference between the full-size CERT and START domain-deleted CERTΔST is the index of lipid-extracting activity via the START domain (Fig. 2).

3.3 CERT START Domain-Mediated Extraction of Ceramide from Phospholipid Membranes: Assay Using SPR

We also developed a novel assay method for the CERT-mediated extraction of natural ceramide from phospholipid membranes, in which the release of lipid molecules from a phospholipid matrix is measured as a loss in weight from the matrix set on an SPR sensor (Fig. 3) [10]. This section describes details of the assay for C$_{16}$-ceramide extraction from a PC/PE matrix by the START domain of CERT. This assay system is widely applicable to other types of LTPs and lipid ligands [*see* **Note 12**].

3.3.1 Preparation of Liposomes

On the day prior to starting the SPR experiment:

1. POPC, POPE, POPS, and C$_{16}$-ceramide are dissolved in chloroform/methanol (1:1 [v/v]) and these lipids are mixed in a glass vial to a final composition of 73.2:18.6:3.20:5.00 (molar ratio) of POPC:POPE:POPS:C$_{16}$-ceramide [*see* **Note 13**].

2. The organic solvents are removed by flash drying a nitrogen gas stream onto the glass vial in order to make a thin lipid cake on the inner wall of the vial [*see* **Note 14**].

3. Residual organic solvents are thoroughly removed by overnight evaporation with a vacuum/freeze dryer [*see* **Note 15**].

4. The syringe holding/heating block of the Mini-extruder is placed in a 37 °C incubator to be pre-warmed before liposome preparation [*see* **Note 16**].

On the day to starting the SPR experiment:

5. After finishing evaporation, 500 μL of pre-warmed (37 °C) vehicle buffer is added to the vial, and the lipid cake is resuspended by gentle back and forth rocking.

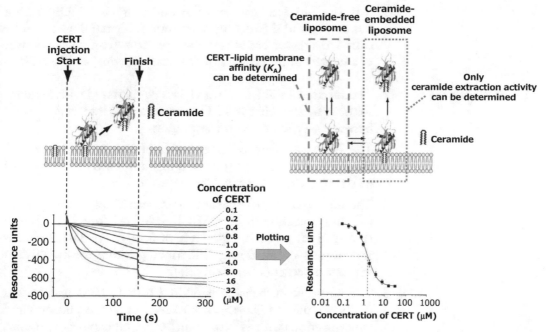

Fig. 3 Schematic diagrams of the SPR-based ceramide extraction assay method. This approach allows the ceramide extraction activity of LTPs and the lipid membrane-LTPs association constant (K_A) to be determined individually. The data presented were published in ref. 10

6. The atmosphere in the vial is substituted with dry nitrogen gas. The vial is tightly capped with a screw plug, shielded from light by wrapping with aluminum foil, and incubated at 37 °C for ~1 h with occasional gentle agitation. Consequently, the lipids are hydrated, and large multi-lamellar lipid vesicles (LMV) are formed [*see* **Note 17**].

7. LMV are disrupted and converted to roughly downsized vesicles by repeatedly freeze-thawing the sample several times (freezing at −80 °C or flash freezing with liquid nitrogen while thawing at 37 °C) [*see* **Note 18**].

8. During this incubation, assemble the Mini-Extruder according to manufacturer's instructions (http://www.avantilipids.com/index.php?option=com_content&view=article&id=185&Itemid=193), and set it on the pre-warmed syringe holding/heating block of the Mini-extruder in the 37 °C incubator until the formation of LMV is completed in **step 7** [*see* **Note 19**].

9. The lipid suspension in the vial is transferred into a gastight syringe. LMV are then converted into small unilamellar vesicles (SUV) by membrane extrusion with the Mini-Extruder in accordance with the manufacturer's instructions (http://www.avantilipids.com/index.php?option=com_content&view=article&id=533&Itemid=297).

10. The SUV suspension is transferred from the gastight syringe into a new empty vial. The atmosphere in the vial is replaced

with dry nitrogen gas, and the vial is capped tightly with a screw plug and shielded by aluminum foil wrapping. It can be stored in a dark place at room temperature (e.g., in the drawer of a bench) until the start of SPR experiments [*see* **Note 20**].

3.3.2 SPR Experiments

1. The surface of the L1 sensor chip is equilibrated with aqueous solution by shedding Milli-Q water at a flow rate of 5 μL/min between several hours and overnight.

2. The L1 sensor chip is further equilibrated by shedding SPR running buffer at a flow rate of 5 μL/min until the baseline of its sensorgram become stable.

3. The surface of the L1 sensor chip is cleaned by a short pulsing (~1 min) injection of 20 mM CHAPS [*see* **Note 21**] solution three to four times at a flow rate of 5 μL/min. The surface is then washed by shedding SPR running buffer for 30–60 min without detergents [*see* **Note 22**].

4. The liposome suspension is diluted to a 1/10 concentration (to be ~1 mM of liposomes with respect to the phospholipid concentration) with SPR running buffer. Liposomes are immobilized on the surface of one channel of the L1 sensor chip by injecting ~500 μL of the diluted liposome suspension at a flow rate of 5 μL/min for 60–90 min until the SPR sensorgram baseline reaches a plateau level. In typical experiments, when the surface of the normal L1 sensor chip is filled with immobilized phospholipid membranes, an increase in resonance units (RUs) up to 8000–10,000 is observed.

5. After further washing the surface of the phospholipid membranes attached on the sensor chip by shedding SPR running buffer for ~30 min, 20 mM NaOH is injected two times in a pulsing manner (~1 min injection period) at a flow rate of 5 μL/min to remove lipid debris nonspecifically bound to the phospholipid membranes.

6. After washing the phospholipid membrane surface by further shedding SPR running buffer for ~30 min, BSA solution (0.1 mg/mL) is injected over the phospholipid membrane surface at a flow rate of 5 μL/min for 30 min of the injection period in order to coat the residual hydrophobic surfaces of the L1 sensor chip. When liposomes are successfully embedded on the L1 sensor chip, a small increase (~20–100 RU) in the sensorgram baseline is observed. Essentially, the same procedures are performed for the chip coated with ceramide-free phospholipid membranes. As a blank reference, the hydrophobic surface of the "noncoated" L1 sensor chip is similarly coated by injecting BSA solution (0.1 mg/mL).

7. Purified CERT for testing (in our published study, we mainly used the START domain of CERT for testing [10]) is injected at a flow rate of 30 μL/min for ~2.5 min until the sensorgram

reaches a plateau level, and the shedding of SPR running buffer is continued for ~2.5 min. When C_{16}-ceramide exists on immobilized phospholipid membranes, a gradual decrease in the SPR sensorgram signal is observed during the injection of CERT along with the release of ceramide from the immobilized phospholipid membrane by CERT (Fig. 3).

8. After finishing the SPR experiments, phospholipid membranes immobilized on the L1 sensor chip are removed by flushing 20 mM CHAPS and followed by an injection of isopropanol in a short pulsing manner (1–2-min injection periods) at a flow rate of 5 μL/min. This causes a marked decrease in the sensorgram baseline to almost baseline levels because of the almost complete removal of lipid membranes from the chip surface.

9. Regarding regeneration, the surface of the L1 sensor chip is washed sufficiently by shedding SPR running buffer at a flow rate of 5 μL/min for 30–60 min typically.

10. The same procedures described above from Subheading 3.3.2, **steps 3–9** are repeated using different concentrations of CERT proteins, which are prepared by serial dilutions in a protein concentration range of 0.1–400 μM (Fig. 3) [see **Note 23**].

11. The ceramide extraction activity of CERT can be determined by plotting the drop width of the final plateau RU values against the concentrations of input CERT (Fig. 3) [10].

12. As a control, essentially the same experiments are also performed for ceramide-free phospholipid membranes formed on the sensor chip. As expected, the CERT-dependent decrease in the SPR sensorgram baseline does not occur in the ceramide-free control experiments [see **Note 24**]. Instead, a slight increase in the sensorgram signal is observed, which represents the binding of CERT to ceramide-free membranes. Thus, the control experiment provides information on the binding affinity (association constant, K_A) of CERT for ceramide-free PC/PE phospholipid membranes [10].

3.3.3 HPLC-MS Analyses

1. In order to collect CERT proteins that have interacted with ceramide-embedded membranes on the L1 sensor chip, the out-flow fluid is retrieved from the SPR equipment in accordance with the SPR manufacturer's instructions [see **Note 25**].

2. Lipids in the out-flow fluid collected are extracted using an organic solvent system [11]: This step is performed at room temperature. Three milliliters of chloroform/methanol (1:2 [v/v]) is added to 0.8 mL of the SPR out-flow fluid in a Pyrex® glass tube [see **Note 26**], mixed by swirling for 3 min [see **Note 27**], and maintained for ~10 min. After the addition of 1 mL of chloroform and 1 mL of 0.5% (w/v) NaCl to the tube, the resultant mixture is mixed well by swirling, and subjected to low-speed centrifugation (1500 × g, 5 min) for phase

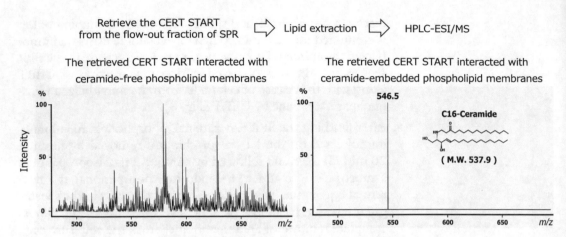

Fig. 4 Mass spectra of the lipid fraction extracted from the CERT START domain in the flow-out fraction of the SPR assay. The retrieved CERT START domains were interacted with ceramide-free (*left panel*) or ceramide-embedded (*right panel*) phospholipid membranes. The data presented were published in Ref. 10

separation. After the removal of the upper phase (aqueous phase) by gentle aspiration or pipetting, the lower phase (organic phase) is retrieved with a glass pipette into a glass tube, and dried under a nitrogen-gas stream in a fume hood.

3. The ceramide molecules are purified and concentrated by performing normal-phase column chromatography with the HPLC system as described below: A Capcell Pak CN column is equilibrated with hexane/ethanol (99:1 [v/v]). The flow rate of the mobile phase is constant (1 mL/min) during all of these chromatographic procedures. Following equilibrium of the solid phase, the lipid sample extracted from the out-flow fluid of the SPR assay is injected into the column, and the solid phase is then washed by shedding the mobile phase for 30 min. After additional washing of the column with hexane/ethanol/2-propanol (98.7:1:0.1 [v/v]) for 10 min, it is followed by running the solvent with a gradually increasing concentration of 2-propanol from 0.1 to 0.3% [v/v] within 10 min. The elution of ceramide molecules is accomplished by isocratic shedding hexane/ethanol/2-propanol (98.7:1:0.3 [v/v]) solution for 10 min, and the eluates are directly introduced into ESI-MS with an ELSD [*see* **Note 28**] (Fig. 4) [10].

4   Notes

1. EDTA and higher concentrations of 2-mercaptoethanol are incompatible with the TALON® metal affinity resin.

2. Moderate-to-high concentrations of salts abrogate nonspecific ionic interactions between molecules. Furthermore, inorganic

phosphate may weakly interact with the cobalt ion of the TALEN® resin, and improve the nonspecific binding of biological components to the resin. Thus, we routinely use 50 mM sodium phosphate buffer (pH 7.0) containing 0.3 M NaCl as the basal buffer in TALEN® resin chromatography.

3. When TCEP is dissolved in aqueous solution, the pH value of the solution markedly decreases to ~2.5. After TCEP is dissolved, readjust pH to the desired value prior to use.

4. An eggplant-shaped glass flask with a similar volume to the glass vials is also acceptable.

5. When the block is heated using a hot plate without an incubator, a thermometer needs to be inserted into the well of the block, and its temperature monitored to ensure that the desired temperature of the block is maintained.

6. Although the purity grade of the nitrogen gas may not be crucial, we recommend >99.5 % purity to prevent the oxidization of lipids.

7. When the recombinant protein is aggregated, the temperature of the overnight culture may be set to 18 °C because hydrophobic interactions, which are often the main cause of protein aggregation, may be weakened under lower temperature conditions.

8. The sample may be precentrifuged at 1400 × g for 15 min in order to precipitate unbroken cells before high-speed centrifugation. At this stage, the size of the pellet needs to be smaller than that of the harvested cells. Large pellets suggest the existence of many unbroken cells. A longer sonication time and/or higher power input may improve the efficiency of lysis. However, stronger sonication conditions often increase the inactivation of proteins of interest.

9. The input and supernatant fractions may be restored in order to check the efficient binding of His_6-tagged CERT. Ideally, His_6-tagged proteins detectable in the input fraction need to be almost absent in the TALON®-unbound supernatant fraction.

10. Several exceptional modifications are made to the assay for the extraction of PC; 0.3 µg of donor vesicles consisting of egg yolk PC, PE, and [*dipalmitoyl*-1-^{14}C] L-α-dipalmitoyl PC (195:65:65, [mol/mol]) are incubated with 600 pmol of purified CERT or CERTΔST.

11. The final concentration of phospholipids becomes 1 mg/mL, and [^{14}C]ceramide is contained at ~0.3 mol% against the total phospholipid amount.

12. This assay system relies on the detection of weight changes in the phospholipid matrix set on the SPR sensor. If LTP molecules continue to bind to the matrix, even after flushing with

an analyte-free buffer, the bound LTP will interfere with the detection of lipid release as a loss in weight from the matrix. Thus, LTPs that transiently interact with the phospholipid matrix are the most suitable for this assay system.

13. At the beginning, stock solutions of 50 mg/mL POPC, POPE, and POPS and 5 mg/mL C_{16}-ceramide are prepared in glass test tubes by dissolving with chloroform/methanol (1:1 [v/v]) and 100 % chloroform, respectively. These individual stock solutions are capped tightly with screw plugs, light-shielded by aluminum foil wrapping, and stored in a dark place at −20 to 30 °C. The lipid mixture to be used is prepared by taking the adequate dose of each stock solution into an empty glass vial as ~10 mM (given that all lipids are dissolved in 0.5 mL of aqueous buffer) of the total lipid concentration. When ceramide-free liposomes are prepared in reference experiments, the C_{16}-ceramide is replaced by POPC to be 78.2:18.6:3.2 (molar ratio) of POPC:POPE:POPS. The lipid composition of the liposome needs to be modified depending on the characteristics/function of the target protein or purpose of the experiments.

14. In a fume food, by rotating the vial around the direction of the long axis while maintaining the incline of the glass vial at ~45°, the lipid solution is spread onto the inside wall of the glass vial. During lipid spreading, the solvent is gradually evaporated by flushing a dry nitrogen gas stream into the thin-layered lipid solution to make a thin lipid cake. This drying process needs to be performed slowly with occasional adjustments of the rotation speed, output flow volume of the nitrogen gas stream, the hitting point of the gas, and the frequency of gas flushing.

15. The mouth of the vial is wrapped with parafilm, and a few pinholes are added using a needle. Residual organic solvent in the vial is evaporated with a vacuum/freeze dryer. During this interval, the vial is light-shielded by wrapping with aluminum foil. The complete removal of residual organic solvent is essential for the formation of stable liposomes.

16. This step can be omitted when a hot plate is used to heat the syringe holding/heating block.

17. Lipid hydration and the formation of LMV need to progress above the gel-liquid crystalline transition temperature. It varies depending on the lipid composition of the desired liposome.

18. Iterative freezing and thawing improve the homogeneity of small lipid particles generated by extrusion; however, liposomes may be formed even if this step is omitted.

19. Although the manufacturer's manual instructs the use of Filter Supports (roughing filters) as "two by two," we routinely use them as "one by one" for economic reasons, and have not encountered any issues specifically caused by this usage format.

However, we identified other issues for the Mini-extruder: if petty clearances exist between the O-rings and polycarbonate membrane, they severely reduce the efficiency of liposome formation. Thus, the following points are considered helpful for successful extrusion: (1) the polycarbonate membrane needs to be placed on one side of the internal support to prevent wrinkles, (2) when the other side of the internal support is placed onto the polycarbonate membrane in order to sandwich the polycarbonate membrane, wrinkles on the polycarbonate membrane need to be prevented, and (3) avoid twisting the polycarbonate membrane between the internal supports when tightening the retainer nut of the outer case.

20. It is preferable to prepare liposomes immediately prior to starting experiments in order for them to be as fresh as possible.

21. 40 mM Octyl- D -glucoside can be used as an alternative to 20 mM CHAPS.

22. Residual detergents interfere with liposome immobilization on the chip, resulting in drifts/gradual decreases in the sensorgram baseline and the disqualification of experiments.

23. The range of protein concentrations needs to be considered depending on the solubility or strength of the lipid extraction activity of individual target LTPs.

24. If CERT gains the ability to extract PC or PE, a drop in the baseline will be observed in the assay using ceramide-free PC/PE membranes. When an amino-acid substitution mutant of the START domain of CERT was assayed in a previous study, a marked drop was observed in the baseline even under ceramide-free conditions, suggesting that the mutation broadened the substrate specificity of CERT, and mutated CERT gained the ability to extract PC and/or PE [10].

25. In order to collect a sufficient amount of ceramide, we repeated this retrieval step ~20 times.

26. Glassware, not plasticware, MUST be used when samples are treated with organic solvents. Polystyrene is not tolerant to chloroform. Although polypropylene is apparently tolerant to chloroform, we do not use any plasticware in this experimental step because organic solvents often extract various ingredients from polypropylene ware, which may interfere with the MS analysis.

27. At this point, the mixture has a one-phase appearance. If phase separation occurs, the addition of several drops (~0.1–0.2 mL) of methanol may ameliorate this issue.

28. The spray and capillary voltages are set to 6 kV and 40 V, respectively. Mass spectra are collected by a scanning range of 520–700 m/z every 0.2 s while maintaining the temperature of the ionization source at 300 °C, and these spectra are then merged.

Acknowledgments

We deeply thank our coworkers for their invaluable contributions to studies concerning CERT. This work was supported by Takeda Science Foundation, and AMED-CREST (to K.H.), and the Ministry of Economy, Trade, and Industry (METI) and the New Energy and Industrial Technology Development Organization (NEDO) of Japan (to S.T.).

References

1. Hanada K, Voelker D (2014) Interorganelle trafficking of lipids: preface for the thematic review series. Traffic 15:889–894

2. Holthuis JC, Menon AK (2014) Lipid landscapes and pipelines in membrane homeostasis. Nature 510:48–57

3. Fukasawa M, Nishijima M, Hanada K (1999) Genetic evidence for ATP-dependent endoplasmic reticulum-to-Golgi apparatus trafficking of ceramide for sphingomyelin synthesis in Chinese hamster ovary cells. J Cell Biol 144:673–685

4. Mesmin B, Bigay J, Moser von Filseck J, Lacas-Gervais S, Drin G, Antonny B (2013) A four-step cycle driven by PI(4)P hydrolysis directs sterol/PI(4)P exchange by the ER-Golgi tether OSBP. Cell 155:830–843

5. Chandler D (2005) Interfaces and the driving force of hydrophobic assembly. Nature 437:640–647

6. Ben-Amotz D, Underwood R (2008) Unraveling water's entropic mysteries: a unified view of nonpolar, polar, and ionic hydration. Acc Chem Res 41:957–967

7. Bankaitis VA, Mousley CJ, Schaaf G (2010) The Sec14 superfamily and mechanisms for crosstalk between lipid metabolism and lipid signaling. Trends Biochem Sci 35:150–160

8. Schulze H, Sandhoff K (2014) Sphingolipids and lysosomal pathologies. Biochim Biophys Acta 1841:799–810

9. Hanada K, Kumagai K, Yasuda S, Miura Y, Kawano M, Fukasawa M, Nishijima M (2003) Molecular machinery for non-vesicular trafficking of ceramide. Nature 426:803–809

10. Sugiki T, Takahashi H, Nagasu M, Hanada K, Shimada I (2010) Real-time assay method of lipid extraction activity. Anal Biochem 399:162–167

11. Bligh EG, Dyer WJ (1959) A rapid method of total lipid extraction and purification. Can J Biochem Physiol 37:911–917

Chapter 7

Quantification of Endogenous Endocannabinoids by LC-MS/MS

Mesut Bilgin and Andrej Shevchenko

Abstract

Here, we describe the LC-MS/MS quantification of 46 molecules representing five major classes of endogenous endocannabinoids and endocannabinoid-related compounds in human blood serum and its lipoprotein fractions.

Key words Endocannabinoids, Endocannabinoid-related compounds, Lipoproteins, LC-MS/MS quantification, Multiple reaction monitoring

1 Introduction

Endocannabinoids (EC) are physiological ligands of CB_1 and CB_2 cannabinoid receptors and important signaling molecules acting in the central and peripheral nervous system. EC are involved in a variety of neurophysiological processes and also implicated in metabolic and cardiovascular diseases [1]. Body fluids (blood plasma) or tissues contain EC in the range of several nM or pg per mg of tissues, respectively [2]. Bona fide EC comprise arachidonic acid or arachidonic alcohol moieties conjugated to a polar head group, such as ethanolamine (anandamide and virodhamine), glycerol (2-arachidonoylglycerol and noladin ether), or dopamine (N-arachidonoyl dopamine). Molecules structurally related to genuine EC (often termed endocannabinoid-related compounds or ERC) comprise other fatty acid/fatty alcohol moieties or different polar head groups, such as amino acids [3]. ERC are less potent ligands of cannabinoid receptors. However, they are involved in numerous biological processes through binding to other receptors or membrane proteins [4].

Because of their low abundance and structural variability, the quantification of endogenous EC and ERC is challenging. EC are typically recovered by the liquid-liquid extraction and quantified

Paul Wood (ed.), *Lipidomics*, Neuromethods, vol. 125,
DOI 10.1007/978-1-4939-6946-3_7, © Springer Science+Business Media LLC 2017

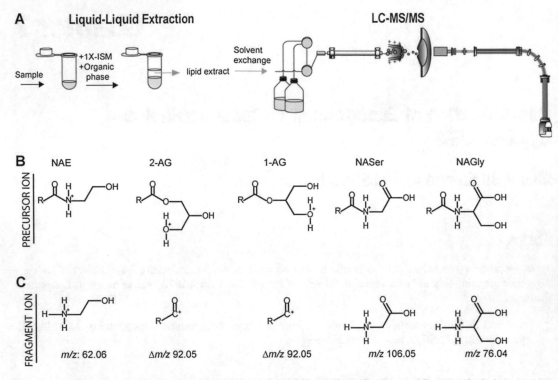

Fig. 1 Quantification of endocannabinoids by MRM LC-MS/MS. (**a**) Quantification workflow that includes one-step liquid–liquid extraction followed by microflow LC-MS/MS analysis using a triple quadrupole mass spectrometer. EC extracts are loaded, concentrated, and desalted on 2 μL reversed phase precolumn and then eluted to the analytical column. (**b**) Chemical structures of protonated molecular ions of major EC classes (proton is shown at the arbitrary position for illustration purposes only). R stands for a hydrocarbon moiety typically comprising 16–22 carbon atoms and 0–6 double bonds. (**c**) Characteristic fragment ions used in the MRM detection of EC. In contrast to NAE, NASer and NAGly 1- and 2- AG are detected using acylium ions of their fatty acid moieties produced by neutral loss of their glycerol head group

by LC-MS/MS in multiple reaction monitoring (MRM) mode. MRM quantification relies on the mass transition between intact protonated precursor ions and polar head group fragments (Fig. 1) [5, 6]. Collision-induced dissociation also yields acylium ions of corresponding fatty acid moieties. However, they are relatively low abundant and their detection is prone to background interference. Nevertheless, they could be used as quantifiers for 1-acyl and 2-acylglycerols and could also serve as useful qualifiers for other ERC support the assignment of their chromatographic peaks [7]. Despite its apparent technical simplicity, MRM LC-MS/MS could be error-prone and lead to inconsistent quantification of EC. Spontaneous isomerization of 2-acylglycerols into 1-acylglycerols; biased losses of some molecules or even EC classes; interference with chemical background and matrix suppression [8] are among most common factors compromising the robustness of measurements. Here, we present an optimized LC-MS/MS method that enables the detection and quantification of 46 molecular species from five major EC and ERC classes in human blood serum and its individual lipoprotein (Lpp) fractions. With minor adjustments, the same method could be used for EC quantification in other model organisms such as *Drosophila* [9].

2 Materials and Standards

Prepare all buffers using LC-MS grade solvents and store them at 4 °C. The internal standard mixture (ISM) should be stored at −80 °C. Perform all operations and store materials (plasticware and glassware) as well as small equipment (pipette, cylinders, centrifuge, and vortex mixer) at 4 °C in a cold room.

Acronyms for endocannabinoid classes: 1-AG, 1-acylglycerol; 2-AG: 2-acylglycerol; NAE: *N*-acylethanolamine; NAGly: *N*-acylglycine; NASer: *N*-acylserine. Individual molecules are annotated by their classes and the number of carbon atoms and double bonds in their fatty acid moieties. Acronyms of synthetic standards additionally contain the number of deuterium atoms *(d)* in their polar head groups.

2.1 Blood Serum and Lipoprotein Fractions

Lipoproteins: VLDL, LDL, HDL, and the plasma fraction devoid of all lipoproteins were obtained by ultracentrifugation basically as described in [10]. Biological materials were stored at −80 °C and kept on ice prior to extraction.

2.2 Mixtures of Internal Standards

1. 10× Internal Standard Mixture (10×-ISM): d4-*N*-acylethanolamine 16:0 (600 nM); d4-*N*-acylethanolamine 18:2 (1915 nM); d8-*N*-acylethanolamine 20:4 (1461 nM); d8-*N*-acylglycine 20:4 (4000 nM); d8-2-acylglycerol 20:4 (3602 nM) and d5-1-acylglycerol 20:4 (4000 nM) in acetonitrile. Store at −80 °C (*see* **Note 1**).

2. 1× Internal Standard Mixture (1×-ISM;): dilute 10×-ISM 10-times with acetonitrile. Store at −80 °C (*see* **Note 2**).

2.3 Extraction Solvents

1. Organic phase: ethyl acetate: *n*-hexane: formic acid (9:1:0.1 *(v/v)*). Mix 18 mL of ethyl acetate, 2 mL of *n*-hexane, and 200 μL of formic acid in 20 mL cylinder (*see* **Note 3**).

2. Water phase: water: formic acid (10:0.1 (v/v)). Mix 20 mL water and 200 μL formic acid (*see* **Note 3**).

3. Resolving buffer: water: acetonitrile: *iso*-propanol: formic acid (6:3.6:0.4:0.1 (v/v)). Mix 12 mL of water, 7.2 mL of acetonitrile, 800 μL of *iso*-PrOH, and 200 μL of formic acid (*see* **Note 3**).

2.4 Eluent for LC-MS/MS Analysis

1. Solvent A: aqueous formic acid (10:0.1 (v/v)). Mix 1 L of water and 1 mL of formic acid.

2. Solvent B: acetonitrile: *iso*-propanol: formic acid (9:1:0.1 (v/v)). Mix 900 mL of acetonitrile, 100 mL *iso*-propanol, and 1 mL formic acid.

3   Methods

Quantification workflow is shown in Fig. 1; all operations should be carried out at 4 °C unless specified otherwise.

3.1 Liquid-Liquid Extraction

1. Mix 750 µL of ethyl acetate-n-hexane-formic acid (9:1:0.1(v/v)), the sample of 500 µL of blood serum (or lipoprotein fractions, or water in control experiments), 50 µL of 1×-ISM, and 12.5 µL of 25 µM PF3845 in 2 mL Eppendorf tube (*see* **Note 4**).

2. Vortex the tube for 30 s and centrifuge at 14,000 g for 10 min.

3. Incubate the tube on dry ice for 10 min and collect the upper (organic) phase into a new 1.5 mL Eppendorf tube (*see* **Note 5**).

4. Dry down the extract in a vacuum centrifuge. Typically, it takes ca 30 min at the set temperature of 37 °C (*see* **Note 6**).

5. Redissolve dried extract in 85 µL of water: acetonitrile: *iso*-propanol: formic acid mixtures (6:3.6:0.4:0.1, (v/v)) and vortex for 30 s (*see* **Note 7**).

6. Clean up the reconstituted extract by centrifuging for 5 min at 14,000 g, collect supernatant, and transfer it into new 1.5 mL Eppendorf tube. Again, spin it down for 5 min at 14,000 g and transfer 85 µL into 300 µL glass vial for LC–MS/MS analysis (*see* **Note 8**).

3.2 LC-MS/MS Analysis

Chromatography is performed on a micro LC 110 system (Agilent Technologies, Santa Clara CA) interfaced online to a triple-quadrupole mass spectrometer TSQ Vantage (Thermo Fisher Scientific, Waltham MA) (*see* **Note 9**).

1. Columns: A trap column C4; inner volume of 2 µL from (Optimize Technologies, Oregon City OR) is directly coupled to 0.5 mm ID × 150 mm analytical column (5 µm Zorbax C18) from Agilent Technologies (*see* **Note 10**). The column was maintained at 40 °C using LC1290 Infinity thermostat (Agilent Technologies) (*see* **Note 11**).

2. Samples were kept at 4 °C and 40 µL was injected using G1377A autosampler (Agilent Technologies) featured with 40 µL loop and Peltier cooling (*see* **Note 12**).

3. Once samples loading was completed and gradient elution was carried out at the flow rate of 20 µL/min (*see* **Note 13**) using the following profile: 0 min, 40% B; 0–5 min, 40% B; 5–7 min, 40–66.4% B; 7–13 min, 66.4–73% B; 13–15 min, 73–95% B; 15–19 min, 95% B; 19–20 min, 95–40% B; 20–24 min; 40% B (isocratic) (*see* **Note 14**).

4. For detecting NAE, NASer and NAGly MRM relies on the mass transition from protonated [M + H]⁺ precursor ions to the

fragments of polar head groups. Contrarily, 1- and 2-acylglycerols are detected via acylium fragments because their polar head group (glycerol) is eliminated as a neutral fragment (Fig. 1). Key MRM settings, such as collision energy (CE) and S-lens voltage, were optimized by the direct infusion of 0.5 μM solutions of standards of each EC / ECR class. Transfer capillary temperature was 275 °C and the ion isolation width of Q1 analytical quadrupole was 0.7 amu (*see* **Note 15**). Mass transitions, CE, and S-lens voltages for most common EC/ERC are provided in Table 1.

Table 1
Mass transitions and instrument settings for MRM analysis of major endocanabinoids on a Vantage triple quadrupole mass spectrometer

Internal standard[a]	Analyte	MRM transition		Collision energy	S-lens
		Precursor ion (*m/z*)	Product ion (*m/z*)	(eV)	(V)
	NAE 16:1	298.3	62.1	16	96
	NAE 16:0	300.3	62.1	16	96
d4-NAE 16:0		304.3	62.1	16	96
	NAE 18:2	324.3	62.1	16	102
	NAE 18:1	326.3	62.1	16	102
	NAE 18:0	328.3	62.1	16	102
d4-NAE 18:2		328.3	66.1	16	102
	NAE 20:5	346.3	62.1	19	108
	NAE 20:4	348.3	62.1	19	108
	NAE 20:3	350.3	62.1	19	108
	NAE 22:6	372.3	62.1	19	108
	NAE 22:5	374.3	62.1	19	108
	NAE 22:4	376.3	62.1	19	108
d4-NAE 20:4		356.3	62.1	19	108
	1-AG 16:1	329.3	237.3	10	78
	1-AG 16:2	331.3	239.3	10	78
	1-AG 18:2	355.3	263.3	11	80
	1-AG 18:1	357.3	265.3	11	80
	1-AG 18:0	359.3	267.3	11	80
	1-AG 20:5	377.3	285.3	15	96
	1-AG 20:4	379.3	287.3	15	96
	1-AG 20:3	381.3	289.3	15	96

(continued)

Table 1
(continued)

Internal standard[a]	Analyte	MRM transition		Collision energy	S-lens
		Precursor ion (*m/z*)	Product ion (*m/z*)	(eV)	(V)
	1-AG 22:6	403.3	311.3	15	96
	1-AG 22:5	405.3	313.3	15	96
	1-AG 22:4	407.3	315.3	15	96
d5-1-AG 20:4		384.3	287.3	15	96
	2-AG 16:1	329.3	237.3	10	78
	2-AG 16:2	331.3	239.3	10	78
	2-AG 18:2	355.3	263.3	11	80
	2-AG 18:1	357.3	265.3	11	80
	2-AG 18:0	359.3	267.3	11	80
	2-AG 20:5	377.3	285.3	15	96
	2-AG 20:4	379.3	287.3	15	96
	2-AG 20:3	381.3	289.3	15	96
	2-AG 22:6	403.3	311.3	15	96
	2-AG 22:5	405.3	313.3	15	96
	2-AG 22:4	407.3	315.3	15	96
d8-2-AG 20:4		387.3	295.3	15	96
	NAGly 16:1	312.3	76.1	13	84
	NAGly 16:0	314.3	76.1	13	84
	NAGly 18:2	338.3	76.1	17	90
	NAGly 18:1	340.3	76.1	17	90
	NAGly 18:0	342.3	76.1	17	90
	NAGly 20:4	362.3	76.1	19	96
	NAGly 22:6	374.3	76.1	19	96
d8-NAGly 20:4		370.4	76.1	19	96
	NASer 16:1	342.3	106.1	15	96
	NASer 16:0	344.3	106.1	15	96
	NASer 18:2	368.3	106.1	17	102
	NASer 18:1	370.3	106.1	17	102
	NASer 18:0	372.3	106.1	17	102
	NASer 20:4	392.3	106.1	19	108
	NASer 22:6	416.3	106.1	19	108
d8-NAGly 20:4		370.4	76.1	19	96

[a]Internal standards used for LC-MS/MS quantification of specified endogenous molecules (shown in rows above the corresponding deuterated standard)

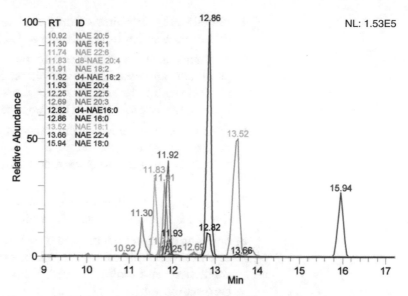

Fig. 2 Overlay of XIC of endogenous NAE from blood serum and the three deuterated standards spiked into the sample prior to EC extraction. XIC were obtained by the method of MRM on a Vantage triple quadrupole mass spectrometer. At the inset RT stands for retention times and ID for the identity of chromatographic peaks

5. The detected molecules were quantified by comparison of the abundances of their extracted-ion-chromatogram (XIC) peaks with peak of corresponding internal standards (Fig. 2; Table 1). If possible, use several internal standards per each quantified class to control for matrix interference (*see* **Note 16**).

4   Notes

1. 10×-ISM is prepared in a 2 mL screw cap glass vial in acetonitrile. It is important to use glass since even at −80 °C long storage of the stock solution in plasticware increases the abundance of peaks of chemical background. Using acetonitrile as a solvent at low (−80 °C) temperature prevents spontaneous isomerization of 2-acylglycerols into 1-acylglycerol. Formic acid also halts the isomerization; however, here it is undesirable because of partial cleavage of EC. It is practical to make fresh aliquots of standards (1×-ISM) by diluting the same concentrated stock (10×-ISM). It reduces biased losses of individual standards and improves the quantification consistency within large batches of samples.

2. 1×-ISM is prepared in a 2 mL screw cap glass vial with acetonitrile and stored at −80 °C.

3. Formic acid prevents spurious isomerization of 2-acylglycerols into 1-acylglycerols.

4. PF3845 is an inhibitor of fatty acid amide hydrolase (FAAH) that otherwise could rapidly cleave EC.

5. Extraction solvent separates into two phases, a water (lower) phase and an organic (upper) phase. Keeping extracted samples on dry ice for 10 min only freezes the water phase and leaves clean organic (upper) phase that is easy to collect.

6. Do not leave samples in the vacuum centrifuge longer than necessary; otherwise, chemical background increases probably due to continuous extraction of plasticizers from tube walls.

7. Keep Eppendorf tubes at 4 °C before adding resolving buffer – this helps to decrease chemical background.

8. Spinning down several times and transferring the sample solution to a new tube minimizes the risk of clogging injection needle, capillaries, and columns and leads to stable reproducible chromatography.

9. Columns with smaller inner diameter decrease the volume of chromatographic peaks and improve the sensitivity. However, we load substantial volumes of organic extracts and therefore nanoflow columns may lack necessary adsorption capacity compared to columns for microflow LC.

10. Using trap column increases loading capacity, supports faster loading and online desalting. Note that, while loading sample onto the precolumn effluent should be directed to waist, but not into the analytical column.

11. High column temperature improves the separation efficiency and symmetry of chromatographic peaks.

12. Samples should be kept at 4 °C to prevent them from drying down. It is also important to experimentally determine the actual injection volume since organic solvents differ by their density and viscosity. Variable injection volume leads to poor quantification consistency.

13. Flow rate should be optimized considering the diameter of particles and the inner diameter of the chromatographic column.

14. The elution gradient starts with 40% of solvent B that matches the composition of resolving buffer. Loading starts at the zero time point and between 0 and 4 min the trap column effluent is directed to waste. After 4 min the trap column is switched in-line to the analytical column.

15. These settings are instrument-dependent.

16. In the absence of matrix interference the quantification of NAE 16:0 and NAE 16:1 using d4-NAE18:2 or d8-NAE 20:4 should be consistent. In case of discordant determinations the

sample preparation protocol should be reexamined to identify the major sources of chemical interference and reduce the quantification bias. It is always recommended to use internal standards most closely resembling chemical structures of quantified analytes.

References

1. Silvestri C, Di Marzo V (2013) The endocannabinoid system in energy homeostasis and the etiopathology of metabolic disorders. Cell Metab 17:475–490

2. Lerner RL, Lutz B, Bindila L (2013) Tricks and tracks in the identification and quantification of endocannabinoids. eLS Doi: 10.1002/9780470015902.a0023407

3. Tortoriello G et al (2013) Targeted lipidomics in *Drosophila melanogaster* identifies novel 2-monoacylglycerols and N-acyl amides. PLoS One 8:e67865

4. Pertwee RG, Howlett AC, Abood ME, Alexander SP, Di Marzo V, Elphick MR, Greasley PJ, Hansen HS, Kunos G, Mackie K, Mechoulam R, Ross RA (2010) International Union of Basic and Clinical Pharmacology. LXXIX. Cannabinoid receptors and their ligands: beyond CB_1 and CB_2. Pharmacol Rev 62:588–631

5. Balvers MG, Verhoeckx KC, Witkamp RF (2009) Development and validation of a quantitative method for the determination of 12 endocannabinoids and related compounds in human plasma using liquid chromatography-tandem mass spectrometry. J Chromatogr B Analyt Technol Biomed Life Sci 877: 1583–1590

6. Balvers MG, Wortelboer HM, Witkamp RF, Verhoeckx KC (2013) Liquid chromatography-tandem mass spectrometry analysis of free and esterified fatty acid N-acylethanolamines in plasma and blood cells. Anal Biochem 434:275–283

7. Fanelli F et al (2012) Estimation of reference intervals of five endocannabinoids and endocannabinoid related compounds in human plasma by two dimensional-LC/MS/MS. J Lipid Res 53:481–493

8. Skonberg C, Artmann A, Cornett C, Hansen SH, Hansen HS (2010) Pitfalls in the sample preparation and analysis of N-acylethanolamines. J Lipid Res 51:3062–3073

9. Khaliullina H, Bilgin M, Sampaio JL, Shevchenko A, Eaton S (2015) Endocannabinoids are conserved inhibitors of the Hedgehog pathway. Proc Natl Acad Sci U S A 112:3415–3420

10. Pietzsch J et al (1995) Very fast ultracentrifugation of serum lipoproteins: influence on lipoprotein separation and composition. Biochim Biophys Acta 1254:77–88

Chapter 8

Lipid Profiling by Supercritical Fluid Chromatography/Mass Spectrometry

Takayuki Yamada and Takeshi Bamba

Abstract

Supercritical fluid chromatography (SFC) is a promising separation technique for comprehensive lipid profiling, or lipidomics. SFC facilitates superior high-throughput separation as compared to liquid chromatography (LC), and can be applied for the simultaneous profiling of diverse lipids with a wide range of polarities due to the physicochemical properties of its mobile phase. To date, analytical methods for fatty acyl, glycerolipid, glycerophospholipid, sphingolipid, sterol lipid, and prenol lipid profiling using SFC coupled to mass spectrometry (MS) have been developed. Hundreds of different molecular species have been identified from biological samples by combining SFC/MS and bioinformatics tools. In addition, SFC can be connected directly with supercritical fluid extraction (SFE), which is suitable for the efficient recovery of hydrophobic compounds that are susceptible to photolytic degradation and oxidation. Online SFE-SFC/MS can be utilized for the high-throughput analysis of the redox status of coenzyme Q_{10} (CoQ_{10}), and phospholipids in dried plasma spots (DPS). These techniques have potential applications in the high-throughput screening of a large number of samples, in biomarker discovery, and in the elucidation of lipid metabolism mechanisms based on the functions of individual lipid molecular species

Key words Lipidomics, Supercritical fluid chromatography, Mass spectrometry, Supercritical fluid extraction

1 Introduction

A supercritical fluid (SCF) is any substance at a temperature and pressure above its critical point (Fig. 1). SCFs have higher diffusivity and lower viscosity than liquids (Table 1); therefore when an SCF is used as the mobile phase during chromatography, efficient separation can be achieved more quickly than in liquid chromatography (LC). In other words, a lower height equivalent to a theoretical plate (HETP) can be obtained at a higher mobile phase flow rate in supercritical fluid chromatography (SFC) than in LC. Supercritical carbon dioxide ($SCCO_2$) has frequently been utilized as a mobile phase for SFC because $SCCO_2$ has low toxicity, is non-flammable, and is chemically inert; it also has a relatively low critical point as compared to other substances. Notably, the use of CO_2 as the principal

Paul Wood (ed.), *Lipidomics*, Neuromethods, vol. 125,
DOI 10.1007/978-1-4939-6946-3_8, © Springer Science+Business Media LLC 2017

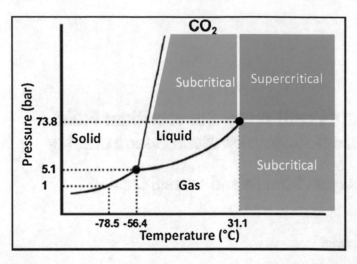

Fig. 1 Pressure-temperature phase diagram of CO_2. Reproduced from Ref. [4] with permission

Table 1
Physical properties for gas, supercritical fluid, and liquid

Type of fluid	Volumetric mass density (g cm^{-3})	Viscosity (cP)	Diffusivity (cm^{-2}·s^{-1})
Gas	10^{-3}	10^{-2}	0.2
Supercritical state	0.5	5×10^{-2}	5×10^{-4}
Liquid	1	1	10^{-5}

Reproduced from Ref [4] with permission

component of the mobile phase decreases the consumption of organic solvents, making SFC a "green" separation technique. In addition, the polarity of SCCO$_2$ is as low as n-hexane [1]; therefore, hydrophobic compounds have good solubility in SCCO$_2$. Moreover, the polarity of the mobile phase in SFC can be adjusted by adding organic solvents such as methanol and acetonitrile, as modifiers or co-solvents. Thus, SFC is suitable for the simultaneous separation of hydrophobic compounds with varying polarities. As such, lipids from several different categories defined by the Lipid Metabolites and Pathways Strategy (LIPID MAPS) consortium [2] have been the main target of supercritical fluid chromatography/mass spectrometry (SFC/MS) analysis [3, 4].

Typically, SFC is divided into two configurations: open tubular capillary column SFC and packed column SFC. Open tubular capillary column SFC was introduced in the 1980s. In this variation, SCCO$_2$ is typically employed as the mobile phase, and the elution strength is controlled by temperature and pressure ramps. Although open tubular capillary column SFC provides high-resolution separation coupled with flame ionization detection (FID), the

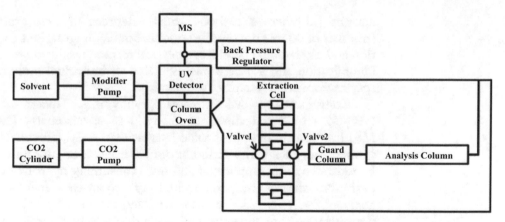

Fig. 2 Schematic diagram of online SFE-SFC/MS system. Reproduced from Ref. [32] with permission

modifiers that can be added to $SCCO_2$ are limited, thus restricting the application to highly hydrophobic compounds. On the other hand, packed column SFC was first introduced as a technique for preparative isolation in the 1990s. The advantage of preparative SFC is the low cost to remove solvents after isolation because CO_2 evaporates naturally at room temperature. Notably, preparative SFC has been utilized for the isolation of chiral compounds. Recently, packed column SFC instruments for mass spectrometric analysis were introduced along with packed columns with various stationary phases, and a number of applications have been reported.

SCFs can also be utilized as extraction solvents. Hydrophobic compounds can be efficiently extracted by supercritical fluid extraction (SFE) due to the diffusivity and hydrophobicity of $SCCO_2$. Notably, SFE has been adopted for the extraction of fatty acids, carotenoids, and fat-soluble vitamins (FSVs) [5–7]. Recently, it was reported that SFE could be applied for hydrophilic metabolites by increasing the ratio of the modifier during extraction [8]. In addition, by coupling SFE and SFC/MS analysis (Fig. 2), a variety of compounds can be extracted and analyzed continuously in a short amount of time without photolytic degradation and oxidation. For example, it has been demonstrated that online SFE-SFC/MS is particularly useful for the analysis of antioxidant compounds.

This chapter focuses on the applications of SCF-related technologies including SFC and SFE in the field of lipidomics. The current status of novel analytical methods based on SFC/MS and online SFE-SFC/MS is discussed, as well as their advantages and drawbacks.

2   SFC/MS Methods for Various Lipid Categories

2.1   *Fatty Acyls*

Fatty acyls include fatty acids and their functional variants such as alcohols, aldehydes, amines, and esters [2]. It is well known that fatty acids impact the regulation of many metabolic processes. There are various types of fatty acids with different carbon chain lengths (number of carbon atoms), degrees of unsaturation (number of double bonds), and double bond configurations (position and *cis/trans*). Therefore, analytical techniques for the precise identification and accurate quantification of individual molecular species from crude mixtures are required.

Conventionally, gas chromatography/mass spectrometry (GC/MS) or liquid chromatography/mass spectrometry (LC/MS) has been utilized to profile fatty acids. In GC/MS analysis, isomeric fatty acids with different double bond configurations can be separated, but complicated and time-consuming hydrolysis and derivatization processes are required prior to analysis. In LC/MS analysis, free fatty acids can be analyzed without hydrolysis or derivatization, but separation of isomeric molecular species with double bonds in different positions using LC is difficult.

SFC complements GC and LC for the separation of fatty acids because it facilitates the separation of individual molecular species with different double bond configurations without the need to prepare derivatized fatty acids [9]. Using an octadecylsilyl (ODS) column with methanol modifier containing 0.1% (v/v) formic acid, isomeric fatty acid species were successfully separated (Fig. 3). Notably, the ODS column was a non-endcapped column, and formic acid was added to methanol in order to decrease the interaction between analytes and free silanol groups in the stationary phase.

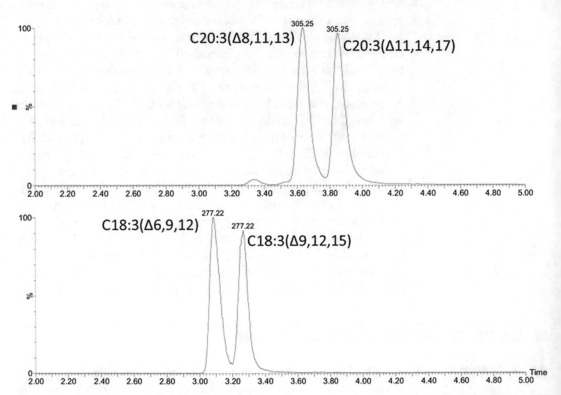

Fig. 3 Extracted ion chromatogram showing the separation of isomeric free fatty acid molecular species based on the chain position of the double bonds. Reproduced from Ref. [9] with permission

Free fatty acids can be analyzed simultaneously with other classes of lipids including glycerolipids, glycerophospholipids, sphingolipids, and sterol lipids [10]. Notably, addition of ammonium acetate instead of ammonium formate or formic acid was favored for the highly sensitive detection of free fatty acids in electrospray ionization mass spectrometry (ESI MS) [11].

Since fatty acids constitute a basic lipid structure, the enhanced separation of fatty acids will contribute to the improved separation of molecular species in other lipid categories.

2.2 Glycerolipids

Glycerolipids are composed of several classes of neutral lipids including acylglycerols and glycosylated acylglycerols [2]. These neutral lipids serve as a source of energy through the β-oxidation of fatty acids, as the hydroxyl groups of glycerol are esterified with fatty acids.

SFC/MS facilitates superior high-throughput analysis of glycerolipids as compared to LC/MS. Triacylglycerols (TAGs), diacylglycerols (DAGs), monoacylglycerols (MAGs), monogalactosyldiacylglycerols (MGDGs), and digalactosyldiacylglycerols (DGDGs) can be analyzed simultaneously with other lipid categories including glycerophospholipids, sphingolipids, and sterol lipids [12, 13]. With reversed-phase columns such as octadecylsilyl (ODS) and phenyl columns, lipid molecular species are separated based on carbon chain length and degree of unsaturation in their fatty acyl moieties [14]. Specifically, regioisomeric TAGs with the same fatty acid moiety in different positions on the glycerol backbone can be resolved using a C30 column with a shorter analysis time than in LC (Fig. 4) [15]. In addition, normal-phase columns

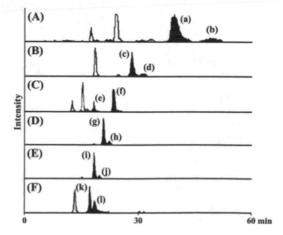

Fig. 4 MRM chromatograms of six regioisomeric TAG pairs in palm and canola oils. (**a**) TAG 18:0/18:1/18:0, (**b**) TAG 18:0/18:0/18:1, (**c**) TAG 18:0/18:1/16:0, (**d**) TAG 18:0/16:0/18:1, (**e**) TAG 18:0/18:3/16:0, (**f**) TAG 18:0/16:0/18:3, (**g**) TAG 16:0/18:1/16:0, (**h**) TAG 16:0/16:0/18:1, (**i**) TAG 16:0/18:2/16:0, (**j**) TAG 16:0/16:0/18:2, (**k**) TAG 16:0/18:3/16:0, (**l**) TAG 16:0/16:0/18:3. Reproduced from [15] with permission

such as silica and 2-ethylpyridine (2-EP) columns can also be utilized for TAG analysis. Since TAGs are highly hydrophobic, they are not retained in normal-phase LC, resulting in poor separation of individual molecular species. On the other hand, TAGs can be separated based on their fatty acyl moieties using a 2-EP column in SFC (Fig. 5) [16]. In this method, TAGs with long carbon chain on their fatty acyl moieties exhibit longer retention times (RTs). This is similar to the case of reversed-phase columns. On the other hand, when the degree of unsaturation in fatty acyl moieties increases, RTs become longer. The elution order is opposite that of reversed-phase columns. In biological systems, the degree of

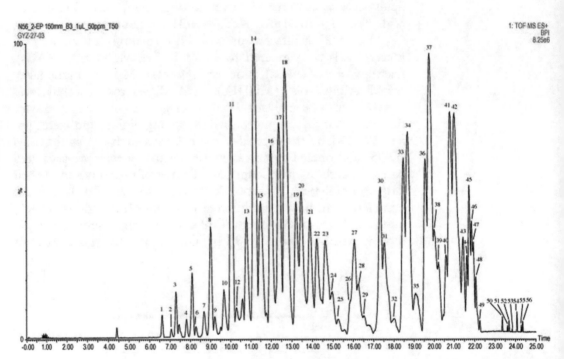

Fig. 5 Base peak ion (BPI) chromatogram of triacylglycerols and diacylglycerols in a lipid extract of cow milk fat. (1) TAG 4:0/6:0/16:0, (2) TAG 6:0/6:0/16:0, (3) TAG 4:0/10:0/14:0, (4) TAG 6:0/10:0/14:0, (5) TAG 4:0/10:0/16:0, (6) TAG 4:0/8:0/18:1, (7) TAG 6:0/10:0/16:0, (8) TAG 4:0/12:0/16:0, (9) TAG 4:0/10:0/18:1, (10) TAG 6:0/12:0/16:0, (11) 4:0/14:0/16:0, (12) TAG 4:0/12:0/18:1, (13) TAG 6:0/14:0/16:0, (14) TAG 4:0/16:0/16:0, (15) TAG 4:0/14:0/18:1, (16) TAG 6:0/16:0/16:0, (17) TAG 4:0/16:0/18:0, (18) TAG 4:0/16:0/18:1, (19) TAG 6:0/18:0/16:0, (20) TAG 6:0/16:0/18:1, (21) TAG 4:0/18:0/18:1, (22) TAG 4:0/18:1/18:1, (23) TAG 16:0/16:0/10:0, (24) TAG 16:0/8:0/18:1, (25) TAG 6:0/18:1/18:1, (26) TAG 10:0/16:0/18:0, (27) TAG 12:0/16:0/16:0, (28) TAG 16:0/10:0/18:1, (29) TAG 18:0/8:0/18:1, (30) TAG 14:0/16:0/16:0, (31) TAG 16:0/12:0/18:1, (32) TAG 10:0/18:1/18:1, (33) TAG 16:0/16:0/16:0, (34) TAG 16:0/14:0/18:1, (35) TAG 16:0/14:0/18:2, (36) TAG 16:0/16:0/18:0, (37) TAG 18:1/16:0/16:0, (38) TAG 18:1/14:0/18:1, (39) TAG 16:0/16:0/18:2, (40) TAG 18:0/16:0/18:0, (41) TAG 18:0/16:0/18:1, (42) TAG 16:0/18:1/18:1, (43) TAG 18:1/16:0/18:2, (44) TAG 18:0/18:0/18:1, (45) TAG 18:1/18:1/18:0, (46) TAG 18:1/18:1/18:1, (47) TAG 18:1/18:0/18:2, (48) TAG 18:1/18:1/18:2, (49) TAG 18:1/18:2/18:2, (50) DAG 16:0/16:0 and/or DAG 14:0/18:0, (51) DAG 16:0/18:0, (52) DAG 16:0/18:1, (53) DAG 14:0/16:0, (54) DAG 16:0/16:0 and/or DAG 14:0/18:0, (55) DAG 16:0/18:0, (56) DAG 16:0/18:1. Reproduced from Ref. [16] with permission

unsaturation usually increases with the increased number of carbon chain length on fatty acyl moieties (the major fatty acid species are FA 18:0, FA 20:4, FA 22:6 among others); therefore, many molecular species in biological samples co-elute when a reversed-phase column is used. As such, 2-EP columns provide favorable selectivity for resolving TAGs from crude mixtures from biological samples.

SFC/MS can also be utilized for the determination of chloro-propanol fatty acid esters, which are structural analogs of acylglyc-erols [17]. SFC/MS facilitates the high-throughput separation of mono-, di-, and tri-esters, which are difficult to separate using LC. This analytical method has potential applications in the detailed profiling of chloropropanol fatty acid esters in edible oils.

Glycerolipids include highly hydrophobic molecular species, and therefore are difficult to separate and elute using reversed-phase LC. Thus, SFC is a powerful tool for the characterization of glycerolipids.

2.3 Glycerophos-pholipids

Glycerophospholipids contain a phosphate group esterified on the glycerol backbone [2]. They are the major components of cell membranes and occasionally function as signal transducers. There are numerous molecular species generated by the combination of substructures, namely the polar head groups and fatty acyl moi-eties. These species exhibit a wide range of polarities depending on

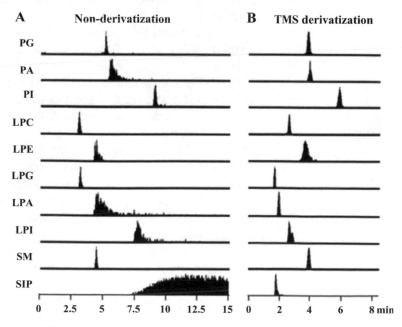

Fig. 6 MRM chromatograms of polar lipids in a mixture of standard samples with and without trimethylsilyl (TMS) derivatization. Reproduced from Ref. [18] with permission

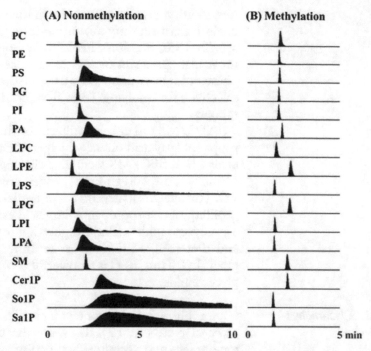

Fig. 7 MRM chromatograms of polar lipids in a mixture of standard samples with and without methylation. Reproduced from Ref. [19] with permission

their constituents. Therefore, analytical methods that allow for the detection of a variety of molecular species are essential for the characterization of glycerophospholipids.

Major classes of glycerophospholipids are consisted of phosphatidylcholine (PC), phosphatidylethanolamine (PE), phosphatidylserine (PS), phosphatidylglycerol (PG), phosphatidylinositol (PI), and phosphatidic acid (PA). Among them, PCs, PEs, and PGs exhibit good peak shapes, while PSs, PIs, and PAs exhibit tailing without derivatization in SFC/MS. The poor peak shapes can be improved by derivatization, primarily via hydroxyl group silylation [18] or phosphate methylation [19] (Figs. 6 and 7). Derivatization of polar functional groups decreases the unfavorable hydrophilic interactions between analytes and the stationary phase of the column, and enhances the solubility in the $SCCO_2$ mobile phase. These derivatization methods also improve the limit of detection.

Two chromatographic separation modes can be applied for glycerophospholipid profiling. With a normal-phase column, glycerophospholipid molecular species are separated based on their polar head groups (Fig. 8); however, with a reversed-phase column, molecular species are resolved based on their fatty acyl moieties (Fig. 9) [12, 20]. In addition, with a polar-embedded ODS column, glycerophospholipids can be separated based on their polar head groups and fatty acid moieties (Fig. 10), thus

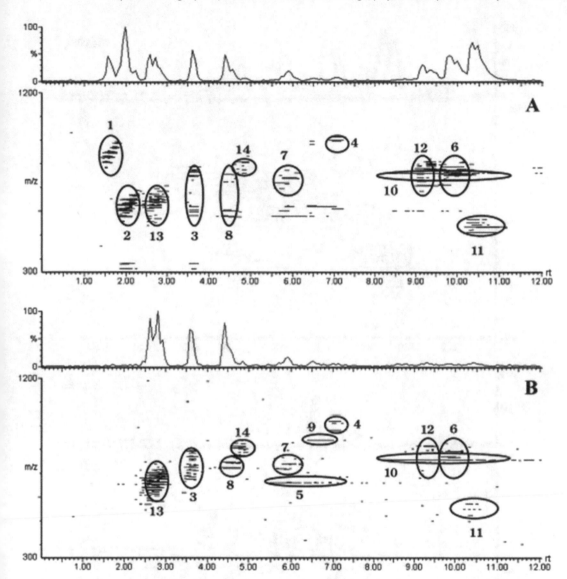

Fig. 8 BPI chromatograms and two-dimensional (2D) displays obtained from SFC/MS analysis of a standard mixture using a cyano column. (*1*) TAG, (*2*) DAG, (*3*) MGDG, (*4*) DGDG, (*5*) PA, (*6*) PC, (*7*) PE, (*8*) PG, (*9*) PI,(*10*) PS, (*11*) LPC, (*12*) SM, (*13*) Cer, and (*14*) cerebroside (CB). Reproduced from Ref. [12] with permission

enabling the precise identification and accurate quantitation of individual molecular species [13].

Also, oxidized glycerophospholipids can be analyzed by SFC/MS. Phospholipids in cell membranes are oxidized by oxidative stress, and as a result, a variety of molecular species with different functional groups such as hydroxides, epoxides, and hydroperoxides are produced, as well as a large number of structural and positional isomers are included. Therefore, analytical techniques for the characterization of such various structural analogs are crucial. Isomeric oxidized PCs have been successfully separated using SFC equipped

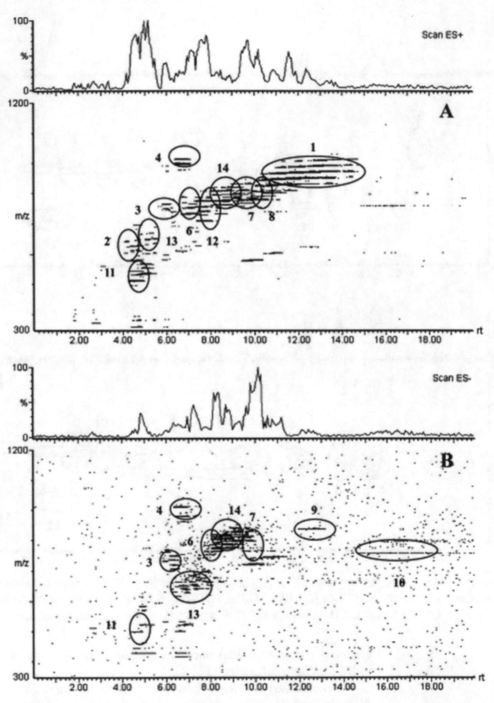

Fig. 9 BPI chromatograms and 2D displays obtained from SFC/MS analysis of a standard mixture using an ODS column. (*1*) TAG, (*2*) DAG, (*3*) MGDG, (*4*) DGDG, (*5*) PA, (*6*) PC, (*7*) PE, (*8*) PG, (*9*) PI, (*10*) PS, (*11*) LPC, (*12*) SM, (*13*) Cer, and (*14*) CB. Reproduced from Ref. [12] with permission

with a 2-EP column [21]. The separation was attributed to the interactions between the positively charged ethylpyridine groups of the stationary phase and negatively charged oxidized PCs.

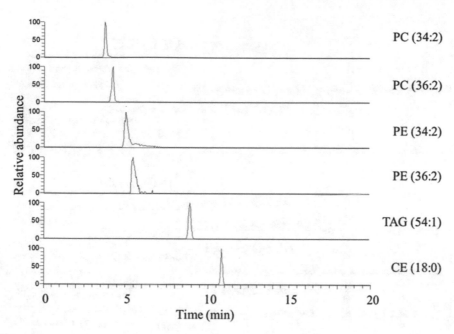

Fig. 10 Extracted ion chromatograms obtained from SFC/MS analysis of a lipid standard mixture using a polar-embedded ODS column. Reproduced from Ref. [13] with permission

Notably, positional isomers of PC epoxides, which could not be resolved in previous studies using LC/MS, can be separated using SFC/MS (Fig. 11), enabling the identification of individual molecular species.

Although SFC was considered to be unsuitable for compounds possessing polar head groups, the use of stationary phases modified with polar functional groups and derivatization methods have negated such concerns. Previous studies demonstrated that SFC/MS can facilitate the simultaneous profiling of glycerophospholipids with a wide range of polarities as well as the highly resolved separation of structural analogs.

2.4 Sphingolipids

Sphingolipids are biomolecules derived from a common backbone composed of long-chain sphingoid bases [2]. The attachment of a variety of N-acyl chains, sugar moieties, and phospho-containing polar head groups enhances the structural diversity and complexity, resulting in several classes: sphingoid bases, ceramides (Cers), phosphosphingolipids, and glycosphingolipids. These species are involved in cellular signaling. Therefore, analytical platforms for the simultaneous determination of sphingolipids are required.

To date, SFC/MS has been applied for the analysis of sphingoid bases, ceramides, phosphosphingolipids, and glycosphingolipids. Similar to glycerophospholipids, several subclasses of phosphosphingolipids including sphingosine-1-phosphate (So1P), sphinganine-1-phosphate (Sa1P), and ceramide-1-phosphate (Cer1P) exhibit

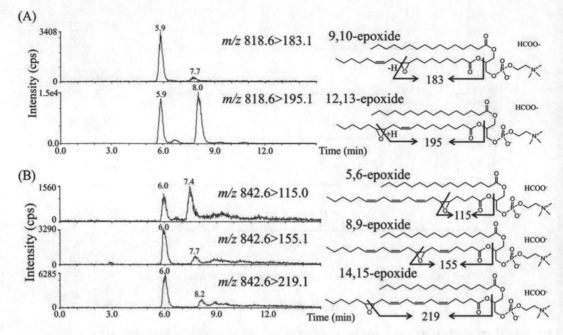

Fig. 11 MRM chromatograms of positional isomers of PC epoxides. Reproduced from Ref. [21] with permission

tailing without derivatization. However, the peak shapes and sensitivity can be improved with silylation or methylation (Figs. 6 and 7) [18, 19]. Additionally, glycosylated molecular species are also important target compounds in the analysis of sphingolipids. Gangliosides have higher polarities than glycerophospholipids and phosphosphingolipids because they possess several saccharide units attached to ceramide. They could be analyzed using SFC coupled to chemical ionization mass spectrometry [22]. Ganglioside subclasses and ceramide heterogeneity were well resolved using a phenyl column and CO_2/methanol mobile phase.

Sphingolipids are an important class of species in lipidome analysis, and several SFC/MS methods have been developed for the simultaneous profiling of major lipid categories in biological systems including glycerolipids, glycerophospholipids, and sterol lipids along with sphingolipids [12, 13, 20]. Notably, SFC can also be applied for the subclass profiling of various sphingolipids possessing phosphates or saccharides.

2.5 Sterol Lipids   SFC/MS has been applied for the analysis of cholesterols and bile acids. These sterol lipids are related to digestion, fat solubilization, and regulation of triglyceride and glucose metabolism.

Free cholesterol (FC) and cholesterol esters (CEs) can be analyzed with other lipids including glycerolipids, glycerophospholipids, and sphingolipids [13, 20]. It was demonstrated that several CE molecular species that were not identified using reversed-phase LC/MS methods could be detected from mouse plasma using SFC/MS [13], indicating that SFC/MS is more sensitive than reversed-phase LC/MS due to the efficient desolvation of the mobile phase (CO_2 and methanol) in the ESI source. Highly hydrophobic metabolites elute only by methanol in SFC equipped with a reversed-phase column, while more hydrophobic solvents such as 2-propanol and acetone, which are not suitable for ESI, are required to elute such compounds in reversed-phase LC.

SFC/MS can be used for the simultaneous profiling of bile acids and their conjugates of varying polarities [23]. When an amide column was used with a methanol/water (95:5, v/v) modifier with 0.2% (w/v) ammonium formate and 0.1% (v/v) formic acid, 25 bile acids eluted with good peak shapes. Furthermore, 24 bile acids were successfully quantified in the analysis of rat serum.

The advantages of SFC have also been demonstrated in the analysis of sterol lipids. SFC facilitates the retention and elution of various metabolites with a wide range of polarities in a single chromatographic run. Moreover, structural analogs including isomers are well resolved within a short time using SFC.

2.6   Prenol Lipids   Carotenoids, which are classified as C40 isoprenoids, are one of the main target analytes in SFC/MS. Carotenoids are hydrophobic antioxidant compounds, and are responsible for oxidative stress. Because of the numerous isomers and structural analogs, chromatographic separation plays a key role in carotenoid profiling [24]. With an end capped polymeric ODS column, isomers were effectively separated, demonstrating that SFC is suitable for high-throughput carotenoid profiling. Another advantage was observed in the detection sensitivity in MS. As compared to conventional reversed-phase LC/MS methods, the developed SFC/MS method exhibited higher sensitivity in ESI-MS due to the difference in the eluent between reversed-phase LC and SFC (Table 2), which enabled the detection of less abundant oxidized carotenoids [25].

SFC/MS was used for the metabolic profiling of β-cryptoxantin (βCX) and its fatty acid esters (βCXFAs), which are carotenoids [26]. Using this method, βCX and βCXFAs from citrus fruits were

Table 2
Comparison of limit of detections (LODs) of β-carotene

Instruments	LOD (fmol)	Mobile phase solvents	Reference
LC-DAD	2000	AcCN/MeOH/CH_2Cl_2	[42]
LC-MS (APCI)	1000	MeOH/MTBE	[43]
LC-MS (ESI)	900	AcCN/MeOH/CH_2Cl_2	[44]
SFC-MS (ESI)	64	$SCCO_2$/MeOH	[24]
SFC-MS/MS (ESI)	0.093	$SCCO_2$/MeOH	[25]

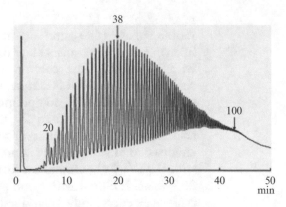

Fig. 12 BPI chromatogram of long-chain polyprenols in *E. ulmoides* leaves. The numbers represent degrees of polymerization for polyprenol homologs. Reproduced from Ref. [29] with permission

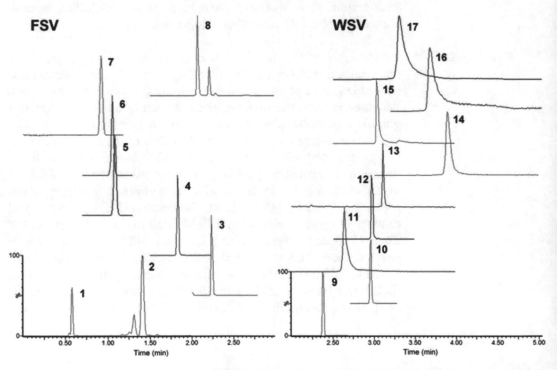

Fig. 13 MRM chromatograms of FSVs and WSVs. (*1*) A acetate, (*2*) A palmitate, (*3*) D2, (*4*) α-tocopherol, (*5*) K2, (*6*) K1, (*7*) α-tocopherol acetate, (*8*) β-carotene, (*9*) nicotinamide, (*10*) Nicotinic acid, (*11*) Pyridoxine, (*12*) D-pantothenic acid, (*13*) Biotin, (*14*) Thiamine, (*15*) Riboflavin, (*16*) B12, (*17*) VC. Reproduced from Ref. [30] with permission

detected with a simple sample preparation method (extraction by acetone) [27].

The utility of SFC/MS was also highlighted in the analysis of polyprenol [28, 29]. The chain length and geometric isomerism of polyprenols from a rubber-producing plant were successfully analyzed using SFC [28]. Additionally, excellent separation of polymer-like long-chain polyprenols was achieved using a phenyl column and tetrahydrofuran modifier (Fig. 12) [29]. SFC/MS exhibited higher chromatographic resolution than LC in the characterization of polyprenols.

Moreover, the simultaneous analysis of fat-soluble vitamins (FSVs) and water-soluble vitamins (WSVs) was achieved by unifying SFC and LC, in "unified chromatography" [30]. In this method, an ODS column without end-capping was utilized, and hydrophobic and hydrophilic vitamins were retained by ODS groups and/or residual silanol groups on the stationary phase. By careful optimization of the analytical conditions, including the composition of the modifier, additives, and column size, all targeted vitamins eluted in a single run (Fig. 13). The gradient elution from almost 100% CO_2 to 100% modifier enabled the simultaneous analysis of FSVs and WSVs with a wide $\log P$ range of -2.11 to 10.12.

Compared with LC/MS, SFC/MS exhibits remarkable advantages in chromatographic resolution and detection sensitivity in the analysis of prenol lipids. Additionally unified chromatography has enabled simultaneous profiling of hydrophobic and hydrophilic vitamins.

3   Online SFE-SFC/MS Methods

There are several practical applications of online SFE-SFC/MS in lipid profiling. Although online SFE-SFC/MS facilitates high-throughput analysis with continuous extraction and chromatographic separation, detailed investigations of the analytical conditions are required during method development because high-efficiency extraction can result in poor chromatographic separation. The details are described below.

A remarkable application is the highly accurate redox status analysis of coenzyme Q_{10} (CoQ_{10}) [31]. CoQ_{10} is an enzyme cofactor classified as a prenol lipid [2], and exists in two forms: a reduced form (ubiquinol-10) and an oxidized form (ubiquinone-10). CoQ_{10} plays an important role in the mitochondrial respiratory chain, and the ratio of those two forms (redox status) of CoQ_{10} is correlated with several diseases. Therefore, accurate analysis of CoQ_{10} redox status is crucial in clinical studies. The reduced form (ubiquinol-10) is recovered with higher efficiency using online SFE-SFC/MS as compared with the conventional method including offline extraction and SFC/MS analysis (Fig. 14). Thus, the reduced form can be extracted without oxidation using online SFE-SFC/MS.

Another application is high-throughput profiling of phospholipids in dried plasma spots (DPS) [32]. DPS testing is a blood sampling method, used for biomarker screening and diagnosis; notably, it is inexpensive and the tested samples are easy to store. In addition, DPS analysis can be combined with online extraction

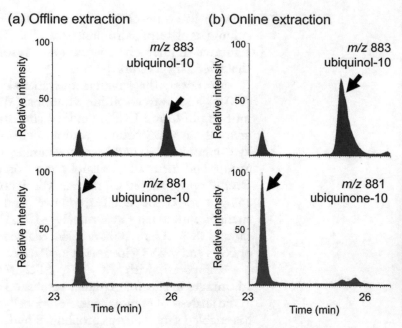

Fig. 14 Extracted ion chromatograms of ubiquinone-10 and ubiquinol-10 in the photosynthesis bacterium, *Rhodobium marinum* A501 obtained by offline extraction (using hexane) or online extraction. Reproduced from Ref. [31] with permission

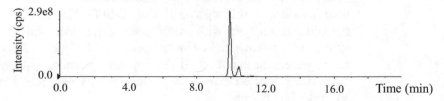

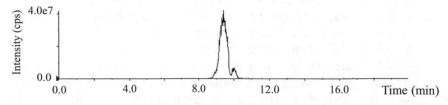

Fig. 15 Extracted ion chromatograms of phospholipid molecular species with different ratios of CO_2 and methanol in extraction

for rapid and simple screening. As such, a practical analytical platform for phospholipid profiling based on online SFE-SFC/MS and DPS was developed [32]. The major difficulty was the optimization of extraction conditions to obtain good chromatographic peak shapes.

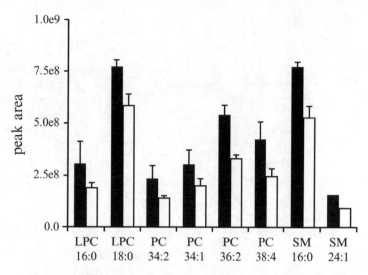

Fig. 16 Comparison of peak areas for each phospholipid molecular species in mouse plasma between SFE (*filled bars*) and LLE (*open bars*). Reproduced from Ref. [32] with permission

The targeted lipid classes were PCs, PEs, and sphingomyelins (SMs) including their lyso-forms; they were extracted with high efficiency when the ratio of the modifier (methanol) was increased during the extraction. However, the extraction conditions produced broad chromatographic peaks (Fig. 15) because the extracted molecular species were not retained in the column during extraction. Therefore, extraction conditions with a lower ratio of the modifier were needed to obtain good peak shapes (Fig. 15). A PC HILIC column was employed because the target phospholipids were strongly retained, and isobaric PCs and SMs could be separated successfully. Using the optimized method, more than 130 molecular species were detected from DPS within 20 min, which included extraction and separation. In addition, the extraction efficiency in SFE is significantly higher than in Bligh and Dyer's method [33], which is a frequently used liquid-liquid extraction (LLE) method (Fig. 16).

As stated above, online SFE-SFC/MS is eminently suitable for the high-throughput analysis of thermally labile and oxidizable hydrophobic compounds. It is expected that the use of online SFE-SFC/MS can be expanded to the profiling of other labile compounds including hydrophilic metabolites.

4   SFC/MS in Lipidome Analysis of Biological Samples

The aforementioned SFC/MS methods have been applied for the lipidomic profiling of biological samples from blood plasma [13, 18], liver [19, 21], brain [20], soybeans [14], and plant leafs [12].

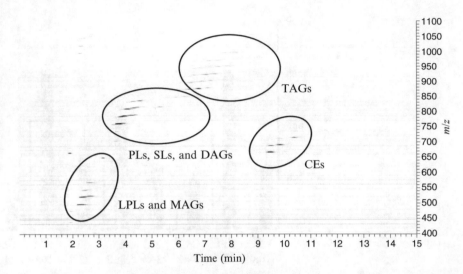

Fig. 17 2D map of lipids in mouse plasma obtained by the SFC/MS analysis of a total lipid extract using a polar-embedded ODS column. Reproduced from Ref. [13] with permission

Several hundreds of molecular species were identified from lipid extracts of each sample. SFC/MS is capable of simultaneous profiling of various lipid classes including fatty acids, acylglycerols, phospholipids, sphingolipids, glycolipids, and cholesterols (Figs. 17 and 18) [12, 13, 20].

Various types of MS can be selected for SFC/MS analysis depending on the objective of the study. For example, a time of flight mass spectrometer (TOF MS) or an Orbitrap Fourier transform mass spectrometer (Orbitrap-FT MS) is suitable for non-targeted or global analysis in which a variety of molecular species are detected using full scanning and product ion scanning with high mass accuracy and resolution [13, 16, 20]. This approach aims at fingerprinting and screening molecular species of interest without any biological hypotheses. For "focused" profiling, precursor ion scanning and neutral scanning in triple quadrupole mass spectrometry (QqQ MS) are employed for the specific detection of preselected target lipid classes [32]. When using multiple reaction monitoring (MRM) mode in QqQ MS, targeted molecular species can be detected with high sensitivity and selectivity, meaning that low-abundance molecular species can be quantified accurately [17, 19, 21, 23, 25, 26].

Data processing software for LC/MS analysis such as Lipid Search [34, 35], LipidBlast [36, 37], MRM-DIFF [38], and MS-DIAL [39] is available for SFC/MS analysis [10, 13] because common mass spectrometers are used in LC/MS and SFC/MS. These bioinformatics tools enable the precise identification and accurate quantification of individual lipid molecular species from complex raw MS data in a short amount of time. By coupling online SFE-SFC/MS analysis and automated lipid identification software, procedures for lipidome analysis including extraction,

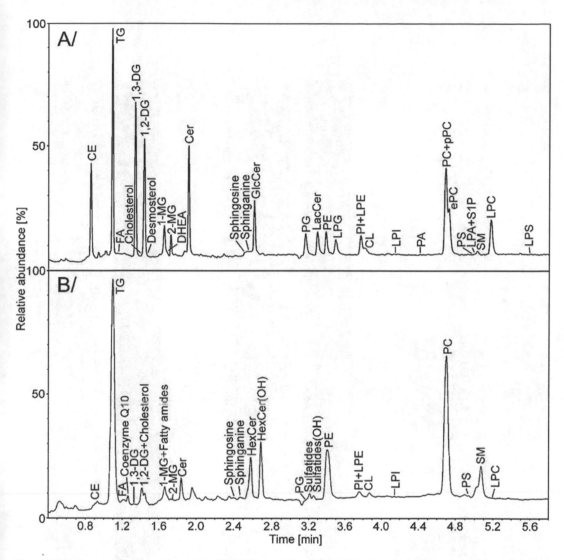

Fig. 18 BPI chromatograms obtained by SFC/MS analysis of (**a**) a lipid standard mixture and (**b**) a total lipid extract of porcine brain using a silica column. Reproduced from Ref. [20] with permission

separation, detection, identification, and quantitation can be carried out within 30 min [32], while it takes several days using conventional methods including LLE, LC/MS analysis, and manual data processing.

A typical application in pathological and physiological research is lipidome analysis of plasma lipoprotein fractions in myocardinal infarction-prone rabbits using the developed SFC/MS platform [10]. In this study, plasma and its lipoprotein collected from normal and Watanabe heritable hyperlipidemic (WHHLMI) rabbits were analyzed individually, and specific alterations in the lipid profiles in each lipoprotein fraction were revealed (Fig. 19). In particular, elevated levels of bioactive lipids including ether-linked phospholipids and ω-6 fatty acid-containing phospholipids in WHHLMI

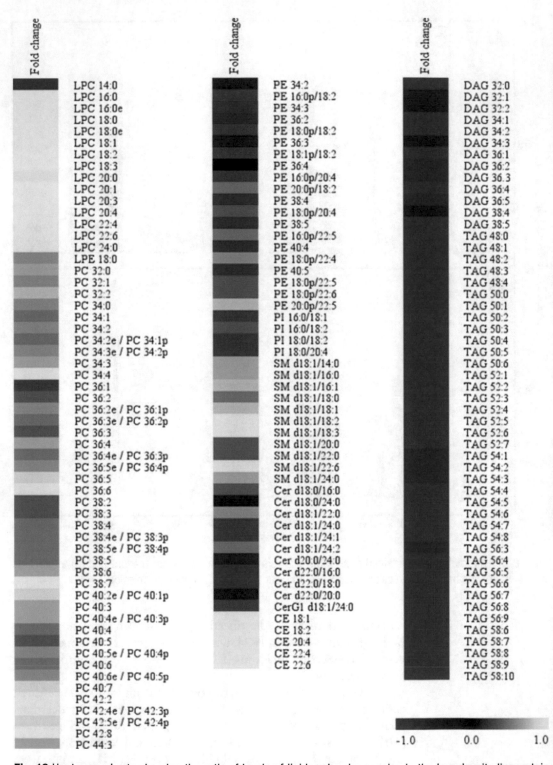

Fig. 19 Heat map charts showing the ratio of levels of lipid molecular species in the low-density lipoprotein (LDL) fraction to those in the very low-density lipoprotein (VLDL) fraction in WHHLMI rabbits. Reproduced from Ref. [10] with permission

rabbits were observed. Therefore, this lipidomics approach based on SFC/MS analysis is expected to be a useful tool for in-depth pathogenic studies on coronary artery diseases.

As mentioned above, SFC/MS is suitable for the high-throughput analysis of a large number of samples, and is expected to be a powerful tool in many research fields including plant, animal, food, and medical studies.

5 Conclusions and Future Perspectives

The applications of SFC/MS in lipidome analysis were reviewed here. SFC/MS provides remarkable benefits including rapid separation of various metabolites with diverse polarities, highly resolved separation of structural analogs, and highly sensitive detection in MS due to the sharp chromatographic peaks and efficient desolvation of the mobile phase. Online coupling of SFC and SFE should also be noted.

However, several issues remain to be solved. First, detailed mechanisms of retention, separation, and elution in SFC should be elucidated and some efforts have been put forth toward that end [40]. Previous studies have revealed great differences in the selectivity between SFC and LC. However, the reason why the difference occurs is still unclear, and there are no established procedures for method development. Second, studies on how to connect columns with different stationary phases in tandem are crucial for the simultaneous analysis of various compounds. Combining several columns is another advantage in SFC, where the column inlet pressure is low due to the low viscosity of the mobile phase. Interestingly, the selectivity varies depending on the order of the connected columns [41]. Therefore, it is expected that structural analogs can be separated by coupling several columns, and the number of analytes retained on the columns will increase. Although it has been demonstrated that various compounds can be analyzed using a single column with mobile phases with a range of polarities, combining several columns would be essential to analyze real lipidomes or metabolomes in a single run.

References

1. Ikushima Y, Saito N, Arai M, Arai K (1991) Solvent polarity parameters of supercritical carbon dioxide as measured by infrared spectroscopy. Bull Chem Soc Jpn 64:2224–2229

2. Fahy E, Cotter D, Sud M, Subramaniam S (2011) Lipid classification, structures and tools. Biochim Biophys Acta 1811:637–647

3. Bamba T, Lee JW, Matsubara A, Fukusaki E (2012) Metabolic profiling of lipids by super-critical fluid chromatography/mass spectrometry. J Chromatogr A 1250:212–219

4. Laboureur L, Ollero M, Touboul D (2015) Lipidomics by supercritical fluid chromatography. Int J Mol Sci 16:13868–13884

5. Cao X, Ito Y (2003) Supercritical fluid extraction of grape seed oil and subsequent separation of free fatty acids by high-speed counter-current chromatography. J Chromatogr A 1021:117–124

6. Careri M, Furlattini L, Mangia A, Musci M, Anklam E, Theobald A, von Holst C (2001) Supercritical fluid extraction for liquid chromatographic determination of carotenoids in spirulina pacifica algae: a chemometric approach. J Chromatogr A 912:61–71

7. Turner C, King JW, Mathiasson L (2001) Supercritical fluid extraction and chromatography for fat-soluble vitamin analysis. J Chromatogr A 936:215–237

8. Matsubara A, Izumi Y, Nishiumi S, Suzuki M, Azuma T, Fukusaki E, Bamba T, Yoshida M (2014) Supercritical fluid extraction as a preparation method for mass spectrometry of dried blood spots. J Chromatogr B 969:199–204

9. Ashraf-Khorassani M, Isaac G, Rainville P, Fountain K, Taylor LT (2015) Study of ultrahigh performance supercritical fluid chromatography to measure free fatty acids with out fatty acid ester preparation. J Chromatogr B 997:45–55

10. Takeda H, Koike T, Izumi Y, Yamada T, Yoshida M, Shiomi M, Fukusaki E, Bamba T (2015) Lipidomic analysis of plasma lipoprotein fractions in myocardial infarction-prone rabbits. J Biosci Bioeng 120:476–482

11. Cajka T, Fiehn O (2016) Increasing lipidomic coverage by selecting optimal mobile-phase modifiers in LC–MS of blood plasma. Metabolomics 12:34

12. Bamba T, Shimonishi N, Matsubara A, Hirata K, Nakazawa Y, Kobayashi A, Fukusaki E (2008) High throughput and exhaustive analysis of diverse lipids by using supercritical fluid chromatography-mass spectrometry for metabolomics. J Biosci Bioeng 105:460–469

13. Yamada T, Uchikata T, Sakamoto S, Yokoi Y, Nishiumi S, Yoshida M, Fukusaki E, Bamba T (2013) Supercritical fluid chromatography/orbitrap mass spectrometry based lipidomics platform coupled with automated lipid identification software for accurate lipid profiling. J Chromatogr A 1301:237–242

14. Lee JW, Uchikata T, Matsubara A, Nakamura T, Fukusaki E, Bamba T (2012) Application of supercritical fluid chromatography/mass spectrometry to lipid profiling of soybean. J Biosci Bioeng 113:262–268

15. Lee JW, Nagai T, Gotoh N, Fukusaki E, Bamba T (2014) Profiling of regioisomeric triacylglycerols in edible oils by supercritical fluid chromatography/tandem mass spectrometry. J Chromatogr B Analyt Technol Biomed Life Sci 966:193–199

16. Zhou Q, Gao B, Zhang X, Xu Y, Shi H, Yu L (2014) Chemical profiling of triacylglycerols and diacylglycerols in cow milk fat by ultra-performance convergence chromatography combined with a quadrupole time-of-flight mass spectrometry. Food Chem 143:199–204

17. Hori K, Matsubara A, Uchikata T, Tsumura K, Fukusaki E, Bamba T (2012) High-throughput and sensitive analysis of 3-monochloropropane-1,2-diol fatty acid esters in edible oils by supercritical fluid chromatography/tandem mass spectrometry. J Chromatogr A 1250:99–104

18. Lee JW, Yamamoto T, Uchikata T, Matsubara A, Fukusaki E, Bamba T (2011) Development of a polar lipid profiling method by supercritical fluid chromatography/mass spectrometry. J Sep Sci 34:3553–3560

19. Lee JW, Nishiumi S, Yoshida M, Fukusaki E, Bamba T (2013) Simultaneous profiling of polar lipids by supercritical fluid chromatography/tandem mass spectrometry with methylation. J Chromatogr A 1279:98–107

20. Lísa M, Holcapek M (2015) High-throughput and comprehensive lipidomic analysis using ultrahigh-performance supercritical fluid chromatography - mass spectrometry. Anal Chem 87:7187–7195

21. Uchikata T, Matsubara A, Nishiumi S, Yoshida M, Fukusaki E, Bamba T (2012) Development of oxidized phosphatidylcholine isomer profiling method using supercritical fluid chromatography/tandem mass spectrometry. J Chromatogr A 1250:205–211

22. Merritt MV, Sheeley DM, Reinhold VN (1991) Characterization of glycosphingolipids by supercritical fluid chromatography-mass spectrometry. Anal Biochem 193:24–34

23. Taguchi K, Fukusaki E, Bamba T (2013) Simultaneous and rapid analysis of bile acids including conjugates by supercritical fluid chromatography coupled to tandem mass spectrometry. J Chromatogr A 1299:103–109

24. Matsubara A, Bamba T, Ishida H, Fukusaki E, Hirata K (2009) Highly sensitive and accurate profiling of carotenoids by supercritical fluid chromatography coupled with mass spectrometry. J Sep Sci 32:1459–1464

25. Matsubara A, Uchikata T, Shinohara M, Nishiumi S, Yoshida M, Fukusaki E, Bamba T (2012) Highly sensitive and rapid profiling method for carotenoids and their epoxidized products using supercritical fluid chromatography coupled with electrospray ionization-triple quadrupole mass spectrometry. J Biosci Bioeng 113:782–787

26. Wada Y, Matsubara A, Uchikata T, Iwasaki Y, Morimoto S, Kan K, Okura T, Fukusaki E, Bamba T (2011) Metabolic profiling of β-cryptoxanthin and its fatty acid esters by supercritical fluid chromatography coupled

with triple quadrupole mass spectrometry. J Sep Sci 34:3546–3552

27. Wada Y, Matsubara A, Uchikata T, Iwasaki Y, Morimoto S, Kan K, Okura T, Fukusaki E, Bamba T (2013) Investigation of β -cryptoxanthin fatty acid ester compositions in citrus fruits Cultivated in Japan. Food Nutr Sci 4:98–104

28. Bamba T, Fukusaki E, Kajiyama S, Ute K, Kitayama T, Kobayashi A (2001) The occurrence of geometric polyprenol isomers in the rubber-producing plant, *Eucommia ulmoides* Oliver. Lipids 36:727–732

29. Bamba T, Fukusaki E, Nakazawa Y, Sato H, Ute K, Kitayama T, Kobayashi A (2003) Analysis of long-chain polyprenols using supercritical fluid chromatography and matrix-assisted laser desorption ionization time-of-flight mass spectrometry. J Chromatogr A 995:203–207

30. Taguchi K, Fukusaki E, Bamba T (2014) Simultaneous analysis for water- and fat-soluble vitamins by a novel single chromatography technique unifying supercritical fluid chromatography and liquid chromatography. J Chromatogr A 1362:270–277

31. Matsubara A, Harada K, Hirata K, Fukusaki E, Bamba T (2012) High-accuracy analysis system for the redox status of coenzyme Q10 by online supercritical fluid extraction-supercritical fluid chromatography/mass spectrometry. J Chromatogr A 1250:76–79

32. Uchikata T, Matsubara A, Fukusaki E, Bamba T (2012) High-throughput phospholipid profiling system based on supercritical fluid extraction-supercritical fluid chromatography/mass spectrometry for dried plasma spot analysis. J Chromatogr A 1250:69–75

33. Bligh EG, Dyer WJ (1959) A rapid method of total lipid extraction and purification. Can J Biochem Physiol 37:911–917

34. Taguchi R, Houjou T, Nakanishi H, Yamazaki T, Ishida M, Imagawa M, Shimizu T (2005) Focused lipidomics by tandem mass spectrometry. J Chromatogr B Analyt Technol Biomed Life Sci 823:26–36

35. Taguchi R, Ishikawa M (2010) Precise and global identification of phospholipid molecular species by an Orbitrap mass spectrometer and automated search engine lipid search. J Chromatogr A 1217:4229–4239

36. Kind T, Liu K-H, Lee DY, DeFelice B, Meissen JK, Fiehn O (2013) LipidBlast *in silico* tandem mass spectrometry database for lipid identification. Nat Methods 10:755–758

37. Kind T, Okazaki Y, Saito K, Fiehn O (2014) LipidBlast templates as flexible tools for creating new *in-silico* tandem mass spectral libraries. Anal Chem 86:11024–11027

38. Tsugawa H, Ohta E, Izumi Y, Ogiwara A, Yukihira D, Bamba T, Fukusaki E, Arita M (2015) MRM-DIFF: data processing strategy for differential analysis in large scale MRM-based lipidomics studies. Front Genet 5:1–15

39. Tsugawa H, Cajka T, Kind T, Ma Y, Higgins B, Ikeda K, Kanazawa M, VanderGheynst J, Fiehn O, Arita M (2015) MS-DIAL: data-independent MS/MS deconvolution for comprehensive metabolome analysis. Nat Methods 12:523–526

40. Lesellier E, West C (2015) The many faces of packed column supercritical fluid chromatography – a critical review. J Chromatogr A 1382:2–46

41. Wang C, Tymiak AA, Zhang Y (2014) Optimization and simulation of tandem column supercritical fluid chromatography separations using column back pressure as a unique parameter. Anal Chem 86:4033–4040

42. Kao TH, Chen CJ, Chen BH (2011) Carotenoid composition in *Rhinacanthus nasutus* (L.) Kurz as determined by HPLC-MS and affected by freeze-drying and hot-air-drying. Analyst 136:3194–3202

43. Albert K, Strohschein S, Lacker T (1999) Separation and identification of various carotenoids by C30 reversed-phase high-performance liquid chromatography coupled to UV and atmospheric pressure chemical ionization mass spectrometric detection. J Chromatogr A 854:37–44

44. Careri M, Elviri L, Mangia A (1999) Liquid chromatography – electrospray mass spectrometry of b-carotene and xanthophylls Validation of the analytical method. J Chromatogr A 854:233–244

Chapter 9

Mass Spectrometric Analysis of Lipid Hydroperoxides

Tânia Melo, Elisabete Maciel, Ana Reis, Pedro Domingues, and M. Rosário M. Domingues

Abstract

Lipids are biomolecules prone to oxidative modifications. In the brain and central nervous system lipids are rich in unsaturated lipids, thus being preferential target for oxidative modifications. Lipid hydroperoxides are the primary products of lipid oxidation, and have been detected in neurodegenerative and neurological disorders. Mass spectrometry-based analytical approaches are nowadays recognized as valuable tools for the identification and quantification of lipid hydroperoxides in specific lipid classes, essential to understand their role in disease onset and progression. Here, we describe the most popular procedure for the lipid extracts from tissues, quantification of total phospholipid content, and quantification of the total content in lipid hydroperoxides. The interpretation of the electrospray mass spectra and tandem mass spectra of lipid hydroperoxides is also explained, exemplified for the case of cardiolipin hydroperoxides molecular species.

Key words Lipid peroxidation, Mass spectrometry, Cardiolipin, Neurodegeneration, Electrospray, FOX II assay, Oxidative lipidomics

1 Introduction

Lipids in brain and central nervous system enriched in polyunsaturated fatty acids are particularly susceptible to oxidation. Many reports describe the increase of lipid peroxidation associated with inflammatory events in neurodegenerative disease (such as Alzheimer, Parkinson) [1–4], autoimmune diseases that affect nervous system (such as multiple sclerosis) [5, 6], or other neurological disorders (such as schizophrenia and depression) [7–9]. This enhances the interest for measuring lipid oxidation in these physiopathological conditions.

Lipid hydroperoxides (LOOH) are the primary products of lipid oxidation in living systems, and can arise from enzymatic catalyzed oxidation, via lipoxygenases, or by non-enzymatic systems by reaction with reactive oxygen species, particularly hydroxyl radical ($^\bullet$OH) [10, 11]. They can decompose in reactive aldehydes such as

Paul Wood (ed.), *Lipidomics*, Neuromethods, vol. 125,
DOI 10.1007/978-1-4939-6946-3_9, © Springer Science+Business Media LLC 2017

malonaldehyde (MDA), thiobarbituric acids (TBA), or 4-hydroxy-nonenal (4-HNE) that are commonly used to estimate lipid oxidation in disease conditions [11]. However, some of these carbonyl compounds can also arise from sugar or protein decomposition making them an unsuitable marker of lipid peroxidation. Monitoring of LOOH in fluids and tissues is a more reliable and suitable marker to estimate on the role of lipid peroxidation in the progression and modulation of neurodegenerative and neurological disorder.

Spectrophotometric methods, such as FOX II assay or ELISA screening kits, can be used to measurement of lipid hydroperoxides, but lack more specific information on the nature of lipid hydroperoxides. Mass spectrometry (MS)-based approaches have been used to identify LOOH in free fatty acids [12–14], phospholipids such as phosphatidylcholines [15–17], phosphatidylethanolamine [18] and cardiolipin (CL) [19–21], glycosphingolipids [22], cholesterol and cholesteryl esters [23, 24], and triglycerides [25].

The development of MS-based approaches assisted (off-line Thin layer chromatography (TLC)-MS and offline high-performance liquid chromatography (HPLC)-MS) or coupled to chromatographic methods (online LC-MS) for the identification and quantification of LOOH, have been used with success to detect specific lipid hydroperoxides molecules in vitro and in vivo. Moreover, lipid hydroperoxides, especially cardiolipin hydroperoxides, have been identified by mass spectrometry-based approaches in substantia nigra and plasma of a model of Parkinson's disease [4], in a mouse model of Alzheimer disease [26], in traumatic brain injury [19], in cerebral ischemia-reperfusion [27], in rat cortical neurons during staurosporine-induced apoptosis [28], and in a model of depression [9].

The identification of lipid hydroperoxides by mass spectrometry requires the interpretation of mass spectra data and the identification of specific fragmentation pathways. Due to the inexistence of commercially available specific standards this fragmentation has been proposed based on finding of biomimetic models using liposomes [16–18, 29, 30], namely typical losses of 32 ($-O_2$) or 34 amu (H_2O_2) that confirm the presence of hydroperoxide moiety, as shown for phosphatidylcholines [16, 17, 29], CL [30], phosphatidylethanolmanie [18], and glycosphingolipids hydroperoxides [22] and the presence of oxidized carboxylate anion product ions $[RCOO+2O]^-$ [18, 19, 21, 30, 31]. The identification of phospholipid hydroperoxides from in vivo experiments is usually based on the separation of phospholipids class by TLC followed by the analyses by MS and MS/MS, as we will report in this chapter for the identification of CL hydroperoxides in brain lipid extracts from an animal models of depression [9]. In brain, CL is particularly rich in PUFA and its location in the mitochondria, the organelle with high production of ROS, makes CL a target for ROS attack. Analysis of CL hydroperoxides started with the preparation of tissue homogenates followed by lipid extraction by the Bligh and Dyer methodology [32] and phospholipid phosphorus determination [33]. The presence of LOOH in the total lipid extract is confirmed by the quantification using FOX assay II.

Phospholipid classes were fractionated by TLC and CL spots that were extracted from silica and the CL extracts were analyzed by electrospray (ESI) mass spectrometry in negative-ion mode. The ESI-MS spectrum of CL species from the brain of the animal models of depression group showed a decrease in relative abundance of the CL $[M-2H]^{2-}$ ions at m/z 737.5 and 749.6, and the increase of ions at m/z 761.7 and 773.6 attributed to the formation of CL hydroxy-hydroperoxy oxidation products (+3O), from the above-described CL species, respectively. The presence of CL hydroperoxides was confirmed by MS^2 and MS^3 analysis using collision-induced dissociation (CID) as an activation technique of doubly charge ions ($[M-2H]^{2-}$).

2 Material

2.1 Extraction of Lipids

2.1.1 Equipment

1. Centrifuge (Mixtasel Centrifuge, JP Selecta, Spain).
2. Vortex (Velp Scientifica, QLABO LDA, Portugal).
3. Pyrex glass cultures tubes with screw-cap (O.D. × L 16 mm × 100 mm, wall thickness 1.8 mm, thread size 15 mm, cap) (Sigma-Aldrich Química, S.L., Portugal).
4. Amber glass vial, screw top, volume 2 mL, thread 8–425, O.D. × H × I.D. 12 mm × 32 mm × 4.6 mm (Supelco, Sigma-Aldrich Química, S.L., Portugal).
5. Screw cap, solid top with PTFE liner 12 mm, for use with 2 mL vial (Supelco, Sigma-Aldrich Química, S.L., Portugal).

2.1.2 Reagents and Supplies

1. Chloroform (HPLC grade, VWR Chemicals, Portugal).
2. Methanol (HPLC grade, Fisher Scientific, UK).
3. Milli-Q water (Millipore Synergy, Merck Millipore, Portugal).
4. 2,6-di-tert-butyl-p-hydroxytoluene (BHT) (Sigma-Aldrich Química, S.L., Portugal).

2.2 Determination of Phospholipid Phosphorus

2.2.1 Equipment

1. Fume hood certified to use $HClO_4$.
2. Heating block to achieve 180 °C (Stuart, Reagent 5, Portugal).
3. Vortex (Velp Scientifica, QLABO LDA, Portugal).
4. Water Bath (Precisterm, JP Selecta, Spain).
5. Microplate reader (Multiskan Go, Thermo Scientific, USA).

6. Pyrex glass cultures tubes with screw-cap (O.D. × L 16 mm × 160 mm, wall thickness 1.8 mm, thread size 15 mm, cap) (Sigma-Aldrich Química, S.L., Portugal).

7. 96-well plates—Microtiter plate in PS with 96-wells flat-bottom 86 × 128 mm (Nuova Aptaca, Italy).

2.2.2 Reagents and Supplies

1. Perchloric acid, $HClO_4$, 70% (Chem-Lab, Belgium).

2. Milli-Q water (Millipore Synergy, Merck Millipore, Portugal).

3. 2.5% (w/v) Ammonium Molybdate (Panreac, Spain)—aqueous solution.

4. 10% (w/v) Ascorbic Acid (VWR Chemicals, Portugal)—aqueous solution.

5. $NaH_2PO_4.2H_2O$ (Riedel-de Haën AG, Switzerland).

2.3 Separation of Phospholipids by TLC

2.3.1 Equipment

1. Thin-layer chromatography developing glass tanks with lids (Sigma-Aldrich Química, S.L., Portugal).

2. TLC silica gel 60 plates (2.5 × 20 cm) with concentration zone (Merck, Germany).

3. Oven (Heraeus, Geprufle Sicherneit, Emílio de Azevedo Campos).

4. Spray for primuline solution (Merck, Germany).

5. UV lamp (Camag, Izasa Scientific, Portugal).

6. Micro Centrifuge 1–14 (Sigma-Aldrich, Reagente 5, Portugal).

7. Ultrasonic Bath (JP Selecta, Spain).

8. Syringe-driven filter unit nonsterile PTFE 0.45 μm × 4 mm, Millex®-H, Millipore (Merck Millipore, Portugal).

9. Syringe 1 mL (Hamilton, Supelco, USA).

10. Amber glass vial, screw top, volume 2 mL, thread 8–425, O.D. × H × I.D. 12 mm × 32 mm × 4.6 mm (Supelco, Sigma-Aldrich Química, S.L., Portugal).

11. Screw cap, solid top with PTFE liner 12 mm, for use with 2 mL vial (Supelco, Sigma-Aldrich Química, S.L., Portugal).

2.3.2 Reagents and Supplies

1. Chloroform (HPLC grade, VWR Chemicals, Portugal).

2. Methanol (HPLC grade, Fisher Scientific, UK).

3. Ethanol (Panreac, Spain).

4. Triethylamine (Acros Organics, Belgium).

5. Milli-Q water (Millipore Synergy, Merck Millipore, Portugal).

6. Acetone (HPLC Grade, Sigma-Aldrich, S.L., Portugal).

7. Primuline (Sigma-Aldrich Química, S.L., Portugal).

8. Phospholipids standards: 1,2-dimyristoyl-*sn*-glycero-3--phosphoethanolamine—$(C_{14:0})_2$-PE; 1,2-dimyristoyl-*sn*-glycero-3-phospho-L-serine (sodium salt)—$(C_{14:0})_2$-PS;

1,2-dimyristoyl-*sn*-glycero-3-phosphocholine—$(C_{14:0})_2$-PC; 1,2-dipalmitoyl-*sn*-glycero-3-phospho-(1′-myo-inositol)—$(C_{16:0})_2$-PI; 1,2-dimyristoyl-*sn*-glycero-3-phosphate (sodium salt)–$(C_{14:0})_2$-PA;   N-oleoyl-D-*erythro*-sphingosylpho-sphorylcholine—(d18:1/18:1)-SM;   1,2-dimyristoyl-*sn*-glycero-3-phospho-(1′-rac-glycerol) (sodium   salt)—$(C_{14:0})_2$-PG,   1-stearoyl-2-hydroxy-sn-glycero-3-phosphocholine—$(C_{18:0})$-LPC; 1,1′,2,2′-tetramy-ristoyl CL (ammonium salt)—$(C_{14:0})_4$-CL (Avanti Polar Lipids, Instruchemie B.V., Netherland).

9. Boric Acid (BDH Chemicals Ltd., England).

2.4   Assay of Lipid Hydroperoxides by FOX II

2.4.1   Equipment

1. Vortex (Velp Scientifica, QLABO LDA, Portugal).
2. Microplate reader (Multiscan Go, Thermo Scientific, USA).
3. Eppendorf tubes (F.L. Medical, Italy).
4. 96-well plates—Microtiter plate in PS with 96-wells flat-bottom 86 × 128mm (Nuova Aptaca, Italy).

2.4.2   Reagents and Supplies

1. Methanol (HPLC grade, Fisher Scientific, UK).
2. Milli-Q water (Millipore Synergy, Merck Millipore, Portugal).
3. Xylenol Orange (Fluka, Sigma-Aldrich Química, S.L., Portugal).
4. 2,6-di-tert-butyl-p-hydroxytoluene (BHT) (Sigma-Aldrich Química, S.L., Portugal).
5. Sulfuric Acid 96% (Farmitalia-Carlo Eba, Laborspirit, Lda, Portugal).
6. $(NH_4)_2Fe(SO_4)_2.6H_2O$ (Iron (II) Ammonium Sulfate Hexahydrate) (Merck, Germany).

2.5   Analysis of Phospholipids by MS

1. Linear ion trap mass spectrometer (ThermoFinnigan, San Jose, CA, USA).
2. Xcalibur operating system (V2.0).

2.5.1   Equipment

2.5.2   Reagents and Supplies

1. Methanol (HPLC grade, Fisher Scientific, UK).
2. Chloroform (HPLC grade, VWR Chemicals, Portugal).
3. 250 μL syringe for direct infusion (Hamilton, Supelco, USA).

3   Methods

3.1   Extraction of Lipids from Tissue Is According the Following Protocol

Total lipids were extracted by Bligh and Dyer procedure [32].

1. Combine 1 mL of brain tissue homogenates with 3.75 mL of a mixture of chloroform and methanol at a ratio 1:2 (v/v).
2. Add 0.1 mg/mL of BHT to the solvents (*see*Note 1).
3. Vortex and keep on ice for 30 min (*see*Note 2).

4. Add 1.25 mL of chloroform and vortex for 1 min.

5. Add 1.25 mL of Milli-Q water and vortex for 1 min.

6. Centrifuge for 5 min at 100 x g.

7. Collect lower phase to a new tube.

8. Add 1.88 mL of chloroform to the aqueous phase and vortex for 1 min.

9. Centrifuge for 5 min at 100 x g.

10. Collect lower phase and join together with the previous one.

11. Evaporate solvent under a stream of nitrogen.

12. Dissolve lipid extracts by adding 300 µL of chloroform and vortex.

3.2 Determination of Phospholipids in Lipid Extracts

Total lipid phosphorus in lipid extracts is determined spectrophotometrically in accordance with the protocol of Bartlett and Lewis [33].

1. Pipet 10 µL of lipid extracts into pyrex glass test tubes.

2. Evaporate solvent to dryness with a nitrogen stream.

3. Add 70% perchloric acid (650 µL) to each tube, and heat to 180 °C for 45 min.

4. After cooling, add 3.3 mL of Milli-Q H_2O to each tube, and vortex.

5. Add 500 µL of 2,5% (w/v) of ammonium molybdate to each tube, and vortex.

6. Add 500 µL of 10% (w/v) of ascorbic acid to each tube, and vortex.

7. Heat to 100 °C for 5 min.

8. After cooling, pipet 200 µL of supernatant to 96-well plates and measure the absorbance of supernatant at 800 nm using a microplate reader (*see* **Note 3**). The total amount of phosphorus of each sample is derived from comparison with the standard curve constructed with known amount of NaH_2PO_4 (0.1–2 µg of phosphorus) (*see* **Note 4**) [34].

3.3 Assay of Lipid Hydroperoxides Determination by FOX II

Lipid hydroperoxides are determined by Ferrous Oxidation-Xylenol Orange (FOX) II assay (*see* **Note 5**) [35].

1. Pipet 50 µL of total lipid extract (*see* **Note 6**) to Eppendorf tubes.

2. Add 950 µL of the FOX II reagent containing 100 µM xylenol orange, 250 µM $(NH_4)_2Fe(SO_4)_2.6H_2O$, 25 mM H_2SO_4, and 4 mM BHT in methanol with water at the ratio of 9:1 (v/v) (*see* **Note 7**).

3. Vortex well and incubate for 30 min, at room temperature, in the dark.

4. Pipet 200 µL of supernatant to 96-well plates and measure the absorbance at 560 nm using a microplate reader.

5. Determine the concentration of the lipid hydroperoxides against H_2O_2 standards with concentrations ranging from 0.0 to 0.4 mM (Fig. 1).

3.4 Separation of Phospholipids by TLC

The phospholipid classes present in lipid extracts are separated by TLC.

1. Wash plates with mixture of chloroform and methanol at the ratio 1:1 (v/v) prior to application and separation of phospholipids by TLC. Let the plates dry in the fume hood for 15 min.

2. Spray plates with 2.3% (w/v) boric acid in ethanol. Let the plates dry in the hood for 15 min.

3. Activate plates by heating at 100 °C for 30 min in the oven to remove all traces of water.

4. Apply lipid extracts (30 µg of phospholipid) (*see* **Note 8**) and authentic phospholipid standards (10 µg of phospholipid) to plate (2.5 × 20 cm). Dry TLC plates with a nitrogen stream to remove the excess of solvents.

5. Develop the plates with a solvent system consisting of chloroform:methanol:H_2O:triethylamine (30:35:7:35, v/v/v/v) (*see* **Note 9**).

6. Reveal the phospholipid spots by spraying the TLC plates with primuline (50 µg/100 mL acetone: water 80/20 (v/v)) (*see* **Note 10**).

7. Visualize the phospholipid spots by using UV lamp ($\lambda = 254$ nm) and identify by comparison with authentic phospholipid standards (Fig. 2a).

8. Scrape the phospholipid spots into pyrex glass test tubes.

9. For phosphorus determination—scrape the phospholipid spots individually into each pyrex glass test tube and estimate the amount of phospholipid phosphorus as described in Sect. 3.3 (*see* **Notes 3** and **4**) (Fig. 2b).

10. For mass spectrometry analysis—scrape the phospholipid spots of each class together into pyrex glass test tubes.

11. Add 2 mL of mixture of chloroform and methanol at a ratio of 1:2 (v/v).

12. Vortex for 2 min.

13. Put samples in ultra-sounds bath for 1 min.

14. Centrifuge at 100 x g for 5 min.

15. Transfer the supernatant to a new glass tube.

16. Wash the pellet by repeating the previous steps.

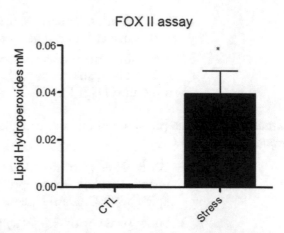

Fig. 1 Lipid hydroperoxide content. Lipid hydroperoxides are the primary products of lipid oxidation and total content in lipid hydroperoxides was assessed using FOX II assay. This method is based on the oxidation of ferrous ions, by hydroperoxides in acid environment, forming ferric ions that are detected using xylenol orange through the formation of a blue–violet complex with an absorption maximum at 560 nm [35]. We found a statistically significant increase of lipid hydroperoxides in the total lipid extracts from brain of animal model of depression, indicating that lipid oxidation occurred in consequence of the stress stimuli. This figure was reproduced from [9] with permission from Elsevier

17. Join the supernatants together and add 1.3 mL of chloroform and 2.4 mL of water.

18. Centrifuge at 100 x g for 5 min.

19. Transfer the organic lower phase to a new glass tube.

20. Evaporate the solvent to dryness with a stream of nitrogen.

21. Resuspend in 100 μL of chloroform and vortex.

22. Filter the sample to an amber vial using syringe-driven filter unit nonsterile PTFE 0.45 μm × 4 mm (*see* **Note 11**).

23. Evaporate the solvent to dryness with a stream of nitrogen and dissolve the phospholipids in 100 μL of chloroform.

3.5 Identification of Cardiolipin and Cardiolipin Hydroperoxides by TLC-ESI-MS and MS²

CL hydroperoxides were identified and characterized by ESI-MS and MS^2 after TLC separation using LXQ with Xcalibur operating system.

1. CL is prepared for ESI-MS analysis by direct infusion for acquisition of ESI mass spectra at a flow rate of 8 μL/min by mixing 40 μL of CL extract in chloroform after TLC in 100 μL of methanol.

2. Operate the electrospray probe at a voltage of 4.7 kV in negative-ion mode.

3. Maintain the source temperature at 275 °C and sheath gas flow rate at 8 units.

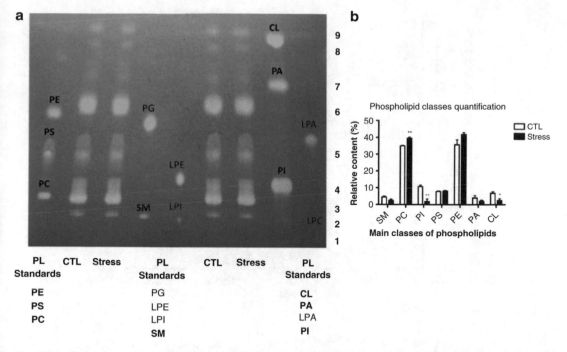

Fig. 2 Thin-layer chromatography (TLC) of total lipid extracts and relative content of phospholipid classes from brain of animal model of depression. (**a**) Typical lipid profile extracted from brain of animal model of depression. Cardiolipin class was identified by comparison with pure and authentic cardiolipin standard applied in the same TLC plate. (**b**) The phospholipid relative content (%) of each class observed in the TLC plate was determined and a significant decrease in cardiolipin levels was observed for brain of animal model of depression. This figure was reproduced from [9] with permission from Elsevier

4. Perform MS analysis using isolation width of 1 m/z, three microscans with maximum injection time 10 ms. For MS^2 analysis, maximum injection time is 100 ms and use collision-induced dissociation (CID) as an activation technique (Q = 0.25).

5. Use full-range zoom (200–2000 m/z) for the MS analysis of CL (*see* **Note 12**). Identify CL either as singly ([M−H]$^-$) or doubly charged ([M−2H]$^{2-}$) ions in the full mass spectrum. Select [M−2H]$^{2-}$ ions to perform MS^2 analysis of CL hydroperoxide derivatives (Fig. 3). Identification was confirmed by the comparison with the fragmentation pattern of a CL hydroperoxide standard—tetra linoleoyl CL hydroperoxide (Fig. 4). Typical fragments are represented in Scheme 1.

4   Notes

1. Lipid extraction has to be performed using BHT and on ice to minimize the risk of oxidation of phospholipids.

2. During the 30 min incubation, vortex samples occasionally (every 10 min).

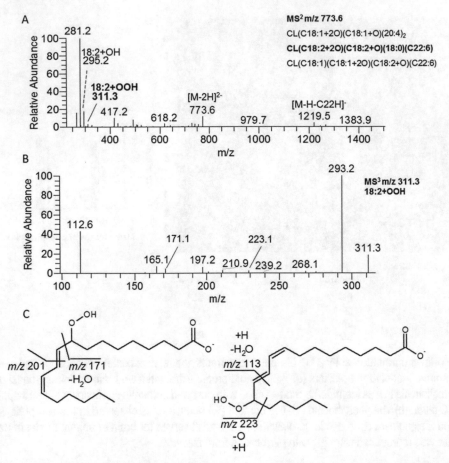

Fig. 3 Characterization of cardiolipin hydroperoxide derived from brain of animal model of depression by ESI-MS. (**a**) MS^2 spectrum of [M−2H]$^{2-}$ ions at m/z 773.6 allowed the identification of carboxylate anion at m/z 311.3 (C18:2+2O) and 295.2 (C18:2+O or/and C18:1+2O−H$_2$O). (**b**) MS^3 analysis of ions at m/z 311.3 showed diagnostic product ions at m/z 293.2, 171.1, and 223.1. Ions at m/z 293.2 are formed due to loss of water from cardiolipin hydroperoxide (CL-OOH) derivative. In fact, loss of water from hydroperoxides during the MS^2 fragmentation is a very common process [30, 36]. (**c**) Ions at m/z 171.1 and 223.1 confirm the presence of hydroperoxide derivative in C-9 and C-13, respectively

3. For TLC samples, scrape the phospholipid spots into pyrex glass test tubes and follow the described protocol. However, before measuring the absorbance of supernatant, transfer 1 mL of sample from pyrex glass tubes to eppendorf tubes, centrifuge for 5 min at 750 x g, and pipet 200 µL of supernatant to 96-well plates and measure the absorbance of supernatant at 800 nm using a microplate reader.

4. The total phospholipid amount can be assessed by multiplying the phosphorus amount value by 25, which is derived by dividing the molecular weight of a phospholipid (average value, ≈770) by the molecular weight of phosphorus, 31.

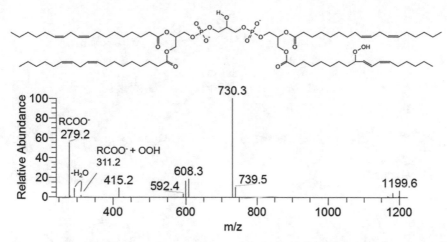

Fig. 4 MS² spectrum of [M−2H]²⁻ ions at *m/z* 739.5 of tetra linoleoyl cardiolipin hydroperoxide (cardiolipin standard) showing the most important diagnostic product ions. The proposed structure for these ions is also represented

5. FOX II assay was developed based on the ability of reduction of ferrous ions (transition metal) by LOOH and reaction of ferrous ions with xylenol orange forming a stable colored complex that can be measured at 560 nm.

6. FOX II assay is performed in total lipid extract instead in TLC lipid spots because at least 50 μg of phospholipid is needed.

7. The FOX II reagent (100 mL) is prepared by dissolving 9.8 mg of $(NH_4)_2Fe(SO_4)_2.6H_2O$ and 139 μL of H_2SO_4 in 5 mL of water. Then, add 88.2 mg of BHT, 7.2 mg of xylenol orange, and 45 mL of methanol. Mix all together and add more 45 mL of methanol and 5 mL of water.

8. Apply three spots of lipid extracts (30 μg of phospholipid) to perform phospholipid phosphorus determination in each phospholipid class and apply four spots of lipid extracts (30 μg of phospholipid) for mass spectrometry analysis.

9. Plates should be developed until 1 cm from the top.

10. Prepare the staining solution by dissolving the primuline powder in water and then add the acetone volume.

11. Before filtering each sample, wash the syringe with 1 mL methanol. Put the filter and wash again with 1 mL methanol. The filter should be used only once.

12. Identification and characterization of lipid hydroperoxides is a complex challenge due to their relatively low abundance and instability. Additionally, overlapping of nonmodified lipid species with the same *m/z* value can occur.

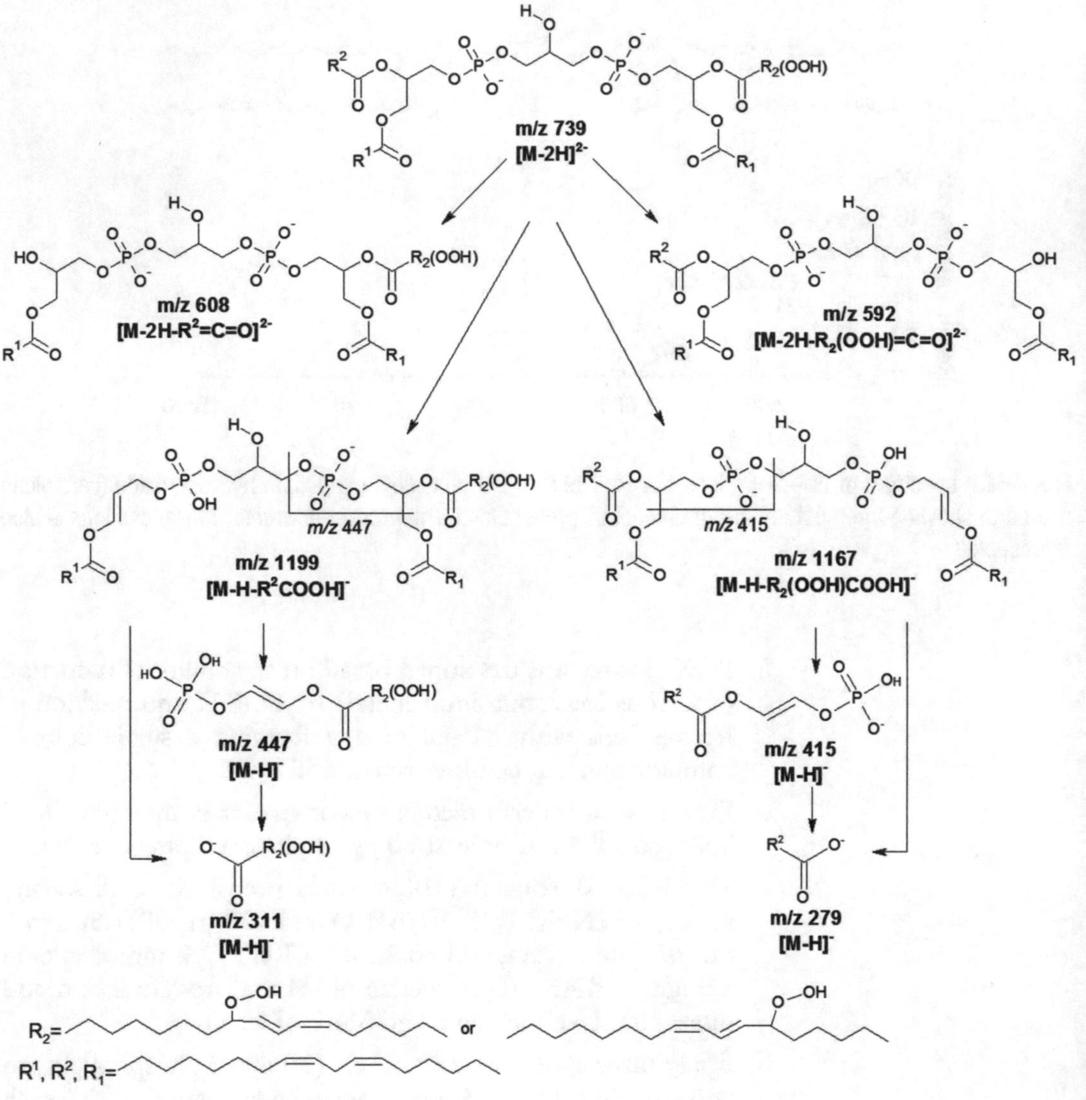

Scheme 1 Fragmentation pathway of [M−2H]$^{2-}$ ions at *m/z* 739.5 of tetra linoleoyl cardiolipin hydroperoxide (cardiolipin standard). The proposed structures for the most important diagnostic product ions were represented

Acknowledgments

Thanks are due to University of Aveiro, Fundação para a Ciência e Tecnologia (FCT, Portugal)/MEC, European Union, QREN, and COMPETE, for funding the QOPNA research unit (PEst-C/QUI/UI0062/2013), and CESAM (UID/AMB/50017/2013), through national founds and where applicable co-financed by the FEDER, within the PT2020 Partnership Agreement, and also to the Portuguese National Mass Spectrometry Network , RNEM, (REDE/1504/REM/2005). Tânia Melo (SFRH/

BD/84691/2012), Elisabete Maciel (SFRH/BPD/104165/2014), and Ana Reis (SFRH/BPD/101916/2014) are grateful to FCT for their grants.

References

1. Galbusera C, Facheris M, Magni F, Galimberti G, Sala G, Tremolada L, Isella V, Guerini FR, Appollonio I, Galli-Kienle M, Ferrarese C (2004) Increased susceptibility to plasma lipid peroxidation in Alzheimer disease patients. Curr Alzheimer Res 1(2)103–109

2. Sanyal J, Bandyopadhyay SK, Banerjee TK, Mukherjee SC, Chakraborty DP, Ray BC, Rao VR (2009) Plasma levels of lipid peroxides in patients with Parkinson's disease. Eur Rev Med Pharmacol Sci 13(2)129–132

3. Montine TJ, Neely MD, Quinn JF, Beal MF, Markesbery WR, Roberts LJ, Morrow JD (2002) Lipid peroxidation in aging brain and Alzheimer's disease. Free Radic Biol Med 33(5)620–626

4. Tyurina YY, Am P, Maciel E, Tyurin VA, Kapralova VI, Winnica DE, Vikulina AS, Domingues MR, McCoy J, Sanders LH, Bayir H, Greenamyre JT, Kagan VE (2015) LC/MS analysis of cardiolipin in substancia nigra and plasma of rotenone-treated rats: Implication for mitochondrial dysfunction in Parkinson's disease. Free Radic Res 49(5)681–691

5. Yousefi B, Ahmadi Y, Ghorbanihaqhjo A, Faghfoori Z, Irannejad VS (2014) Serum arsenic and lipid peroxidation levels in patients with multiple sclerosis. Biol Trace Elem Res 158(3)276–279

6. Tully M, Zheng L, Shi R (2014) Acrolein detection: potential theranostic utility in multiple sclerosis and spinal cord injury. Expert Rev Neurother 14(6)679–685

7. Dietrich-Muszalska A, Kontek B (2010) Lipid peroxidation in patients with schizophrenia. Physchiatry Clin Neurosci 64(5)469–475

8. Hatch J, Andreazza A, Olowoyeye O, Rezin GT, Moody A, Goldtein BI (2015) Cardiovascular and psychiatric characteristics associated with oxidative stress markers among adolescents with bipolar disorder. J Psychosom Res 79(3)222–227

9. Faria R, Santana MM, Aveleira CA, Simões C, Maciel E, Melo T, Santinha D, Oliveira MM, Peixoto F, Domingues P, Cavadas C, Domingues MR (2014) Alterations in phospholipidomic profile in brain of mouse model of depression induced by chronic unpredictable stress. Neuroscience 273:1–11

10. Reis A, Spickett CM (2012) Chemistry of phospholipid oxidation. Biochim Biophys Acta 1818(10)2374–2387

11. Ayala A, Muñoz MF, Argüelles S (2014) Lipid peroxidation: production, metabolism, and signaling mechanisms of malondialdehyde and 4-hydroxy-2-nonenal. Oxid Med Cell Longev 2014:360438

12. MacMillan DK, Murphy RC (1995) Analysis of lipid hydroperoxides and long-chain conjugated keto acids by negative ion electrospray mass spectrometry. J Am Soc Mass Spectrom 6(12)1190–1201

13. Hall LM, Murphy RC (1998) Electrospray mass spectrometric analysis of 5-hydroperoxy and 5-hydroxyeicosatetraenoic acids generated by lipid peroxidation of red blood cell ghost phospholipids. J Am Soc Mass Spectrom 9(5)527–532

14. Nakamura T, Bratton DL, Murphy RC (1997) Analysis of epoxyeicosatrienoic and monohydroxyeicosatetraenoic acids esterified to phospholipids in human red blood cells by electrospray tandem mass spectrometry. J Mass Spectrom 32(8)888–896

15. Adachi J, Matsushita S, Yoshioka N, Funae R, Fujita T, Higuchi S, Ueno Y (2004) Plasma phosphatidylcholine hydroperoxide as a new marker of oxidative stress in alcoholic patients. J Lipid Res 45(5)967–971

16. Reis A, Domingues MR, Amado FM, Ferrer-Correia AJ, Domingues P (2007) Radical peroxidation of palmitoyl-linoleoyl-glycerophosphocholine liposomes: Identification of long-chain oxidized products by liquid chromatography-tandem mass spectrometry. J Chromatogr B Analyt Technol Biomed Life Sci 855(2)186–199

17. Reis A, Domingues P, Domingues MR (2013) Structural motifs in primary oxidation products of palmitoyl-arachidonoyl-phosphatidylcholines by LC-MS/MS. J Mass Spectrom 48(11)1207–1216

18. Melo T, Santos N, Lopes D, Alves E, Maciel E, Faustino MA, Tomé JP, Neves MG, Almeida A, Domingues P, Segundo MA, Domingues MR (2013) Photosensitized oxidation of phosphatidylethanolamine monitored by electrospray

tandem mass spectrometry. J Mass Spectrom 48(12)1357–1365

19. Bayir H, Tyurin VA, Tyurina YY, Viner R, Ritov V, Amoscato AA, Zhao Q, Zhang XJ, Janesko-Feldman KL, Alexander H, Basova LV, Clark RS, Kochanek PM, Kagan VE (2007) Selective early cardiolipin peroxidation after traumatic brain injury: an oxidative lipidomics analysis. Ann Neurol 62(2)154–169

20. Tyurin VA, Tyurina YY, Jung MY, Tungekar MA, Wasserloss KJ, Bayir H, Greenberger JS, Kochanek PM, Shvedova AA, Pitt B, Kagan VE (2009) Mass-spectrometric analysis of hydroperoxy- and hydroxyl-derivatives of cardiolipin and phosphatidylserine in cells and tissues induced by pro-apoptotic and pro-inflammatory stimuli. J Chromatogr B Analyt Technol Biomed Life Sci 877(26)2863–2872

21. Tyurina YY, Tyurin VA, Kanyar AM, Kapralova VI, Wasserloss K, Li J, Mosher M, Wright L, Wipf P, Watkins S, Pitt BR, Kagan VE (2010) Oxidative lipidomics of hyperoxic acute lung injury: mass spectrometric characterization of cardiolipin and phosphatidylserine peroxidation. Am J Physiol Lung Cell Mol Physiol 299(1)L73–L85

22. Couto D, Santinha D, Melo T, Ferreira-Fernandes E, Videira RA, Campos A, Fardilha M, Domingues P, Domingues MR (2015) Glycosphingolipids and oxidative stress: Evaluation of hydroxyl radical oxidation of galactosyl and lactosylceramides using mass spectrometry. Chem Phys Lipids 191:106–114

23. Adachi J, Kudo R, Ueno Y, Hunter R, Rajendram R, Want E, Preedy VR (2001) Heart 7-hydroperoxycholesterol and oxysterols are elevated in chronically ethanol-fed rats. J Nutr 131(11)2916–2920

24. Harkewicz R, Hartvigsen K, Almazan F, Dennis EA, Witztum JL, Miller YI (2008) Cholesteryl ester hydroperoxides are biologically active components of minimally oxidized low density lipoprotein. J Biol Chem 283(16)10241–10251

25. Giuffrida F, Destaillats F, Skibsted LH, Dionisi F (2004) Structural analysis of hydroperoxy- and epoxy-triacylglycerols by liquid chromatography mass spectrometry. Chem Phys Lipids 131(1)41–49

26. Monteiro-Cardoso VF, Oliveira MM, Melo T, Domingues MR, Moreira PI, Ferreiro E, Peixoto F, Videira RA (2015) Cardiolipin profile changes are associated to the early synaptic mitochondrial dysfunction in Alzheimer's disease. J Alzheimers Dis 43(4)1375–1392

27. Ji J, Baart S, Vikulina AS, Clark RS, Anthonymuthu TS, Tyurin VA, Du L, St Croix CM, Tyurina YY, Lewis J, Skoda EM, Kline AE, Kochanek PM, Wipf P, Kagan VE, Bayir H (2015) Deciphering of mitochondrial cardiolipin oxidative signaling in cerebral ischemia-reperfusion. J Cereb Blood Flow Metab 35(2)319–328

28. Tyurin VA, Tyurina YY, Feng W, Mnuskin A, Jiang J, Tang M, Zhang X, Zhao Q, Kochanek PM, Clark RS, Bayir H, Kagan VE (2008) Mass-spectrometric characterization of phospholipids and their primary peroxidation products in rat cortical neurons during staurosporine-induced apoptosis. J Neurochem 107(6)1612–1633

29. Reis A, Domingues P, Ferrer-Correia AJ, Domingues MR (2004) Tandem mass spectrometry of intact oxidation products of diacylphosphatidylcholines: evidence for the occurrence of the oxidation of the phosphocholine head and differentiation of isomers. J Mass Spectrom 39(12)1513–1522

30. Maciel E, Domingues P, Domingues MR (2011) Liquid Chromatography/tandem mass spectrometry analysis of long-chain oxidation products of cardiolipin induced by the hydroxyl radical. Rapid Commun Mass Spectrom 25(2)316–326

31. Hall LM, Murphy RC (1998) Analysis of stable oxidized molecular species of glycerophospholipids following treatment of red blood cell ghosts with t-butylhydroperoxide. Anal Biochem 258(2)184–194

32. Bligh EG, Dyer WG (1959) A rapid method of total lipid extraction and purification. Can J Biochem Physiol 37(8)911–917

33. Bartlett E, Lewis D (1970) Spectrophotometric determination of phosphate esters in the presence and absence of orthophosphate. Anal Biochem 36(1)159–167

34. Hanahan DJ (1997) Choline-containing phospholipids: diacyl-, alkylacyl-, and alkenylacylcholine phosphoglycerides and sphingomyelin. In: A guide to phospholipid chemistry. Oxford University Press, New York, pp 61–130

35. Jiang ZY, Woollard AC, Wolff SP (1991) Lipid hydroperoxide measurement by oxidation of Fe^{2+} in the presence of xylenol orange. Comparison with the TBA assay and an iodometric method. Lipids 26(10)853–856

36. Kim J, Minkler PE, Salomon RG, Anderson VE, Hoppel CL (2011) Cardiolipin: characterization of distinct oxidized molecular species. J Lipid Res 52(1)125–135

Chapter 10

Mass Spectrometric Determination of Fatty Aldehydes Exemplified by Monitoring the Oxidative Degradation of (2*E*)-Hexadecenal in HepG2 Cell Lysates

Corinna Neuber, Fabian Schumacher, Erich Gulbins, and Burkhard Kleuser

Abstract

Within the last few decades, liquid chromatography-mass spectrometry (LC-MS) has become a preferred method for manifold issues in analytical biosciences, given its high selectivity and sensitivity. However, the analysis of fatty aldehydes, which are important components of cell metabolism, remains challenging. Usually, chemical derivatization prior to MS detection is required to enhance ionization efficiency. In this regard, the coupling of fatty aldehydes to hydrazines like 2,4-dinitrophenylhydrazine (DNPH) is a common approach. Additionally, hydrazones readily react with fatty aldehydes to form stable derivatives, which can be easily separated using high-performance liquid chromatography (HPLC) and subsequently detected by MS. Here, we exemplarily present the quantification of the long-chain fatty aldehyde (2*E*)-hexadecenal, a break-down product of the bioactive lipid sphingosine 1-phosphate (S1P), after derivatization with 2-diphenylacetyl-1,3-indandione-1-hydrazone (DAIH) via isotope-dilution HPLC-electrospray ionization-quadrupole/time-of-flight (ESI-QTOF) MS. Moreover, we show that the addition of *N*-(3-dimethylaminopropyl)-*N*'-ethylcarbodiimide hydrochloride (EDC hydrochloride) as a coupling agent allows for simultaneous determination of fatty aldehydes and fatty acids as DAIH derivatives. Taking advantage of this, we describe in detail how to monitor the degradation of (2*E*)-hexadecenal and the concurrent formation of its oxidation product (2*E*)-hexadecenoic acid in lysates of human hepatoblastoma (HepG2) cells within this chapter.

Key words (2*E*)-hexadecenal, (2*E*)-hexadecenoic acid, Sphingosine 1-phosphate, Derivatization, DAIH, EDC, Isotope-dilution, HPLC-ESI-QTOF

1 Introduction

Sensitive and selective detection of low-abundant compounds is a fundamental challenge in biological sciences. For quite some time, LC-MS has become an indispensable tool to face this challenge. While MS has undergone huge technological developments within the last decades that ensured an improvement of the overall analytical performance, analyte-related limitations remained. Due to the

Paul Wood (ed.), *Lipidomics*, Neuromethods, vol. 125,
DOI 10.1007/978-1-4939-6946-3_10, © Springer Science+Business Media LLC 2017

chemical nature of certain compound classes, insufficient ion yields are achieved by ESI frequently used in LC-MS setups. So it is the case for fatty aldehydes. These lipid metabolites with a chain length of at least 12 carbon atoms exert a variety of biological functions. In *Lepidoptera* species some fatty aldehydes are characterized as sex pheromones [1]. Further, retinal (retinaldehyde) is essential for animal vision [2]. But, in general, fatty aldehydes are central products of various cellular pathways. For instance, in plasmalogens, a specific group of glycerophospholipids, fatty aldehydes are bound to a glycerol backbone via a vinyl ether linkage at *sn*-1 position. The interest in fatty aldehyde analysis emerged in the last decades, since the role of plasmalogens in disorders like Alzheimer's disease [3], ischemia [4], and multiple sclerosis [5] became obvious. An impaired oxidative metabolism of fatty aldehydes is also associated with a pathologic situation known as the Sjögren-Larsson syndrome [6].

A general analytical challenge while subjecting fatty aldehydes to mass spectrometric detection is the low ionization efficiency and insufficient fragmentation. In this regard, fatty aldehydes with conjugated diene systems serve as exceptions. Using electron ionization (EI), which is a hard ionization technique, and MS coupling, characteristic fragmentation patterns and high sensitivity were observed [7]. However, also unconjugated fatty aldehydes are susceptible for sensitive MS detection due to the reactivity of the carbonyl function that enables coupling to nucleophilic derivatization agents [8]. First attempts involved the conversion of fatty aldehydes to thermally stable dimethyl acetal (DMA) derivatives in an acidified methanolic milieu [9]. Coupling of gas chromatography (GC) with EI-MS consequently showed characteristic fragment ions for DMA derivatives to be $[M-31]^+$ due to the loss of a methoxy group and ions with a mass-to-charge ratio (m/z) of 75 (due to $[CH(OCH_3)_2]^+$). Another prominent derivatization reagent is O-(2,3,4,5,6-pentafluorobenzyl)hydroxylamine hydrochloride (PFBHA). Under mild buffered conditions at room temperature stable pentafluorobenzyl oximes (PFBOs) are formed. Analysis of the oximes is done using GC and GC-MS [10].

Today, soft ionization techniques, like ESI, atmospheric pressure chemical ionization (APCI), and matrix-assisted laser desorption ionization (MALDI), are most common in mass spectrometry laboratories. In particular, ESI has become the most popular ionization technique. It has the advantage of LC compatibility and generation of protonated or deprotonated molecular ions of polar to moderate polar small molecules or high-molecular weight compounds. Consequently, manifold chemical derivatization reagents for the LC-MS detection of aldehydes were established. Among those, hydrazines are preferred coupling reagents for fatty aldehydes. For instance, the prominent DNPH (Brady's reagent) readily reacts with carbonylic compounds at room temperature in acidic media by forming stable 2,4-dinitrophenylhydrazones [11, 12]. Those hydrazones are usually separated by reversed phase (RP)-HPLC and characterized via ultraviolet (UV) or fluorescence detection or by mass spectrometry after negative ionization. Also hydrazones itselves can be effective derivatization reagents for fatty aldehydes. For instance, it was shown that DAIH readily reacts with fatty aldehydes, exemplarily shown for the derivatization of (2E)-hexadecenal and detection via HPLC-ESI-QTOF MS [13]. As a special feature, the addition of EDC as a coupling agent allows for simultaneous detection of fatty aldehydes and fatty acids as DAIH derivatives [14]. This finding is of importance for the conduction of oxidative metabolism studies of fatty aldehydes. In general, aldehydes are reactive compounds that readily form Schiff base products with amino groups of cellular macromolecules like proteins [15, 16] and DNA [17, 18]. Substituted aldehydes, like α-chlorofatty aldehydes [19], or α,β-unsaturated aldehydes (e.g., 4-hydroxy-*trans*-2,3-nonenal [20]) can be further attacked by cellular nucleophiles (e.g., glutathione) away from the carbonyl function. To prevent cell damage, aldehydes are rapidly detoxified for instance by oxidation via intracellular aldehyde dehydrogenases (ALDH) into their corresponding acids. It was shown that (2E)-hexadecenal, formed after irreversible degradation of the sphingolipid S1P by S1P lyase, can be further oxidized to (2E)-hexadecenoic acid by the fatty aldehyde dehydrogenase ALDH3A2 (known as FALDH) [21]. The protocol presented herein is suitable to answer the question, to what extent (2E)-hexadecenal is endogenously oxidized to (2E)-hexadecenoic acid in vitro. Therefore, lysed HepG2 cells were incubated with (2E)-hexadecenal over a period of 60 min under physiological conditions. At defined time points samples were taken and extracted using a modified protocol published by Püttmann et al. [22]. For quantification of the monitored fatty aldehyde and its corresponding fatty acid, we used the isotope-dilution approach, which is considered to be the gold standard for unambiguous quantification of various types of analytes in biological samples [23]. Hence, stable-isotope labeled (2E)-hexadecenal (d_5) and hexadecanoic acid (d_5) were added as internal standards. The dry extracts were derivatized using DAIH and EDC prior to mass spectrometric analysis using HPLC-ESI-QTOF.

2   Materials

All chemicals and solvents used were of highest available purity or LC-MS grade, respectively. All glassware for extraction was rinsed with ethanol prior to use to avoid contamination with exogenous sources of fatty acids (*see* **Note 1**).

2.1 Cell Lysis and Incubation with (2E)-Hexadecenal

1. Cell pellets contained 2×10^6 HepG2 cells in each Eppendorf tube and were stored at −80 °C.

2. Potassium phosphate buffer: 100 mM, pH 7.4.

3. Ultrasonic homogenizer: Sonoplus UW2070 (Bandelin, Berlin, Germany).

4. Water bath: Grant JB1 (Grant Instruments, Cambridge, UK).

5. Vortex mixer: Vortex Genie 2 (Scientific Industries, New York, USA).

6. (2E)-hexadecenal (Avanti Polar Lipids, Alabaster, USA): Prepare a 408 µM working solution in ethanol. Store at −20 °C (*see* **Note 2**).

7. For investigation of the oxidative degradation of (2E)-hexadecenal (d_5) (results not presented herein; for additional information see Neuber et al. [14]): Prepare a 408 µM working solution of (2E)-hexadecenal (d_5) (Avanti Polar Lipids, Alabaster, USA) in acetonitrile. Store at −20 °C (*see* **Notes 2** and **3**).

8. 1.5 mL screw-capped glass vials.

2.2 Extraction

1. Internal standard mixture A: Prepare a mixture of (2E)-hexadecenal (d_5) and hexadecanoic acid (d_5) (Sigma Aldrich, Schnelldorf, Germany) with a concentration of 20 µM each in ethanol. Store at −20 °C (*see* **Note 2**).

2. Alternative internal standard mixture B for monitoring the oxidative degradation of (2E)-hexadecenal (d_5): Prepare a 20 µM mixture of pentadecanal (Sigma Aldrich, Schnelldorf, Germany) and heptadecanoic acid (Sigma Aldrich, Schnelldorf, Germany) in ethanol. Store at −20 °C (*see* **Note 2**).

3. Solvents used for extraction: Water, *n*-heptane, Dole solution (isopropanol/*n*-heptane/2 M phosphoric acid (40:10:1, *v/v/v*)). Store at 4 °C.

4. 1.5 mL screw-capped glass vials and 400 µL glass micro-inserts.

5. Vortex mixer: Vortex Genie 2 (Scientific Industries, New York, USA).

6. Centrifuge: Biofuge fresco (Heraeus Holding GmbH, Hanau, Germany).

7. Savant SpeedVac Concentrator (Thermo Fisher Scientific, Dreieich, Germany).

2.3 Derivatization

For a set of 50 samples prepare the following reagents in glass tubes:

1. DAIH solution: Dissolve 1.5 mg DAIH (Sigma Aldrich, Schnelldorf, Germany) in 2 mL acetonitrile. Sonicate for 15 min in a water bath before adding 0.5 mL ethanol. Store on ice until use (*see* **Note 4**).

2. EDC solution: Dissolve 119.9 mg EDC hydrochloride (Sigma Aldrich, Schnelldorf, Germany) in 2.5 mL 3% ethanolic pyridine (dilute 75 µL pyridine with 2425 µL ethanol) to yield a 250 mM solution. Store on ice until use (*see* **Note 5**).

3. Further required solvents: Ethanol, water.

4. Ultrasonic water bath: Sonorex RK100H (Bandelin, Berlin, Germany).

5. Thermomixer comfort and 2.0 mL Safe-lock tubes (Eppendorf AG, Hamburg, Germany).

6. 1.5 mL screw-capped glass vials.

2.4 Mass Spectrometry

1. Agilent 1200 LC system coupled to an Agilent 6530 QTOF MS (both from Waldbronn, Germany) (*see* **Note 6**).

2. Eluent A: Water; eluent B: Methanol containing 0.01% ammonium hydroxide.

3. Separation column: ZORBAX Eclipse XDB-C18 (4.6 × 50 mm, 1.8 µm) (Agilent, Waldbronn, Germany).

4. Guard column: ZORBAX Extend-C18 (2.1 × 12.5 mm, 5 µm) (Agilent, Waldbronn, Germany).

3 Methods

Cell material was stored on ice during sample preparation.

Cell Lysis and Incubation with (2E)-Hexadecenal

1. Resuspend a cell pellet in 500 µL ice-cold potassium phosphate buffer.

2. Sonicate the cell solution three times for 20 s on ice, while waiting 30 s between each sonication step (*see* **Note 7**).

3. Transfer the whole cell lysate into a 1.5 mL screw-capped glass vial.

4. Preheat the whole cell lysate for 5 min in a 37 °C water bath before adding 10 µL (2*E*)-hexadecenal working solution (or add 10 µL of (2*E*)-hexadecenal (d$_5$) working solution for studying the oxidative metabolism of the labeled aldehyde instead) and vortex briefly.

5. Before starting the incubation with the aldehyde at 37 °C, 2 µL of a test substance (e.g., modifier of the fatty aldehyde-to-acid metabolism such as an inhibitor of aldehyde dehydrogenases) in ethanol can be added to the sample (if so, add 2 µL ethanol also to control samples). Otherwise add 2 µL of potassium phosphate buffer.

6. Over a period of 60 min and at defined time points, 10 µL of the cell lysates are taken for extraction.

3.1 Extraction of (2E)-Hexadecenal and (2E)-Hexadecenoic Acid

Extraction is done on ice. All solutions are stored on ice prior to use.

1. 10 μL of incubated cell lysate is placed in a 1.5 mL glass vial and filled up with 90 μL water.

2. Immediately add 500 μL Dole solution and 5 μL internal standard mixture A for incubation with (2E)-hexadecenal or 5 μL alternative internal standard mixture B for incubation with (2E)-hexadecenal (d$_5$) (see **Note 8**).

3. Vortex 30 s before adding 200 μL n-heptane and 300 μL water. Again thoroughly vortex for 30 s.

4. It follows a 10 min incubation step on ice.

5. The extracts are centrifuged at $400 \times g$ (4 °C) for 5 min.

6. 200 μL of the upper organic phase is transferred into a 400 μL micro-insert and evaporated to dryness under reduced pressure.

3.2 Derivatization of (2E)-Hexadecenal and (2E)-Hexadecenoic Acid

1. Dried extracts were dissolved in a mixture of 25 μL ethanol, 50 μL DAIH solution, 50 μL EDC solution, and 75 μL water (see **Note 9**).

2. Place the micro-inserts containing the redissolved extracts in 2.0 mL Safe-lock Eppendorf tubes half-filled with water and preheated to 80 °C. Lock the tubes.

3. Heat the derivatization mixture for 15 min at 80 °C. The underlying chemical reactions are depicted in Fig. 1. Let the samples cool down for at least 1 h at room temperature (see **Note 10**). Place the micro-inserts in new 1.5 mL screw-capped glass vials. The derivatized samples should be stored at 4 °C until analysis (see **Note 11**).

3.3 Mass Spectrometry

1. For analyzing levels of derivatized (2E)-hexadecenal and (2E)-hexadecenoic acid using HPLC-ESI-QTOF MS, the following instrumental conditions were chosen.

3.3.1 LC Conditions

1. Separation column: C$_{18}$ phase, the column temperature was set to 30 °C.

2. Injection volume: 10 μL, the temperature of the autosampler was set to 4 °C.

3. Flow rate: 0.9 mL/min.

4. Eluent A: Water; eluent B: Methanol containing 0.01% ammonium hydroxide (Tables 1 and 2).

3.3.2 Settings of the MS/ MS Detector

1. Precursor ions, fragment ions, and optimized collision energies for the selected reaction monitoring (SRM) of DAIH derivatives of fatty aldehydes and acids investigated are as follows Table 3.

   The product ion mass spectra of derivatized (2E)-hexadecenal and (2E)-hexadecenoic acid are presented in Fig. 2.

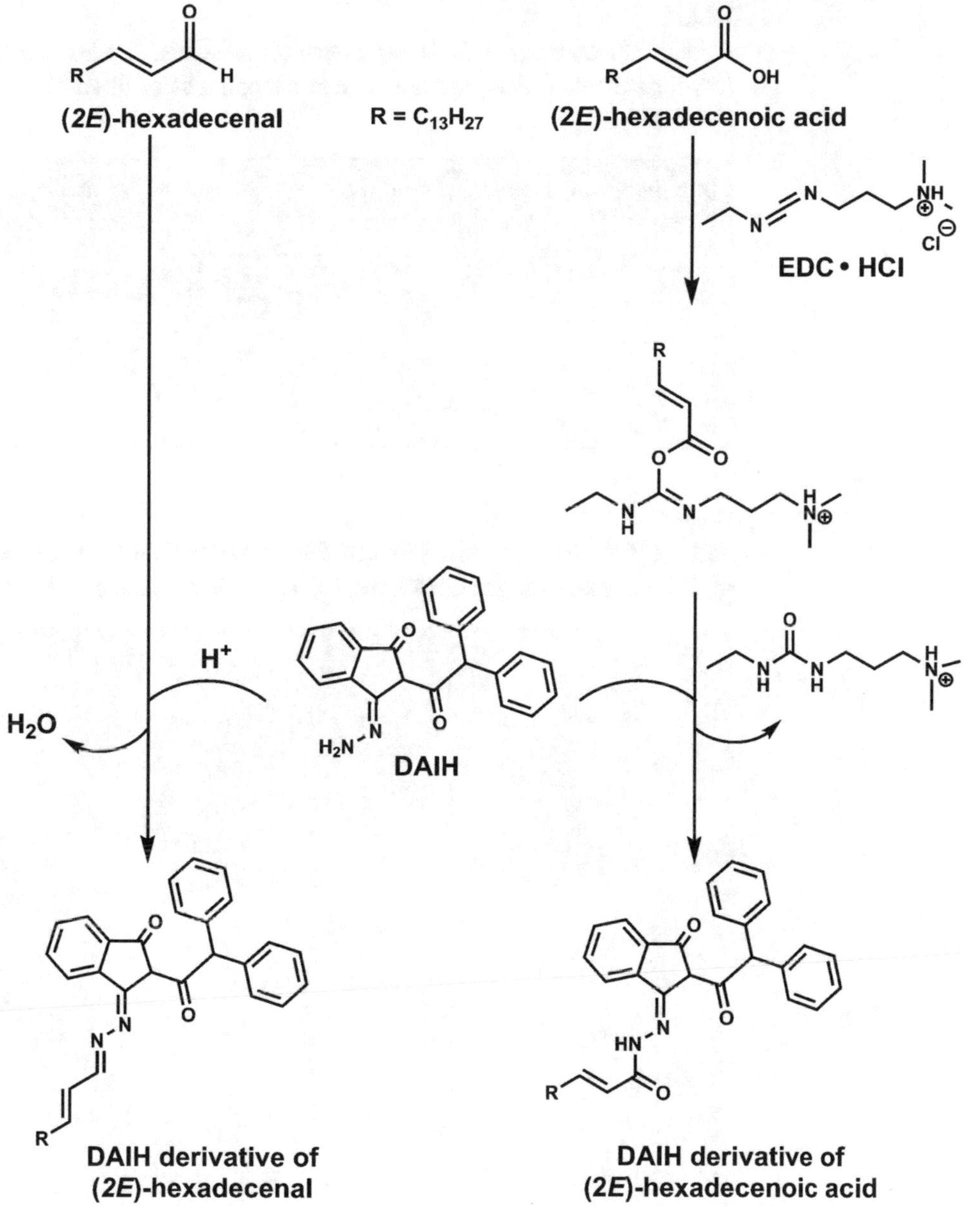

Fig. 1 Chemical reactions underlying the simultaneous derivatization of the fatty aldehyde (2*E*)-hexadecenal and its oxidized metabolite (2*E*)-hexadecenoic acid using DAIH and EDC

2. (2*E*)-hexadecenal and (2*E*)-hexadecenal (d₅) co-eluted at 8.3 min. (2*E*)-hexadecenoic acid and hexadecanoic acid (d₅) co-eluted at 1.9 min. A representative SRM chromatogram showing the signals of derivatized (2*E*)-hexadecenal, (2*E*)-hexadecenoic acid, and their corresponding internal standards is given in Fig. 3.

3. For quantification, peak areas of the analytes were normalized to those of their corresponding internal standards. The obtained

Table 1
Gradient elution program for LC separation of derivatized (2*E*)-hexadecenal, (2*E*)-hexadecenoic acid, and their stable-isotope labeled internal standards

Time (min)	Eluent A (%)	Eluent B (%)
0	15	85
3	0	100
12	0	100
12.1	15	85
18	15	85

Table 2
Settings of the AJS (Agilent Jet Stream) ESI source for simultaneous detection of DAIH derivatives of (2*E*)-hexadecenal and (2*E*)-hexadecenoic acid

Parameter	Setting/value
Ionization mode	Negative (ESI-)
Drying gas temperature	350 °C
Drying gas flow	9 L/min of nitrogen
Sheath gas temperature	400 °C
Sheath gas flow	9 L/min of nitrogen
Nebulizer pressure	22 psi
Capillary voltage	4000 V
Nozzle voltage	250 V
Fragmentor voltage	200 V
Skimmer voltage	65 V
Octopol voltage	750 V

ratio was divided by the slope of an external calibration curve (*see* **Note 12**).

4. A typical time course of the oxidative degradation of (2*E*)-hexadecenal and the concurrent formation of (2*E*)-hexadecenoic acid in HepG2 cell lysates is shown in Fig. 4.

4   Notes

1. Do not use siliconized material. The contamination of samples with hexadecanoic acid is enormous and impedes accurate fatty acid analysis.

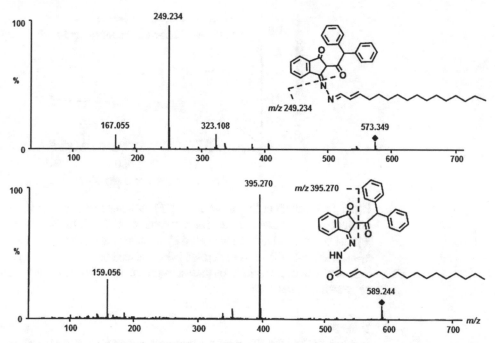

Fig. 2 Product ion mass spectra of derivatized (2*E*)-hexadecenal (*upper* spectrum) and (2*E*)-hexadecenoic acid (*lower* spectrum). The chemical structures of the analyzed DAIH derivatives and the fragmentations utilized for SRM are given in the insets. MS/MS parameters: ESI-, collision energies: as given in Table 3

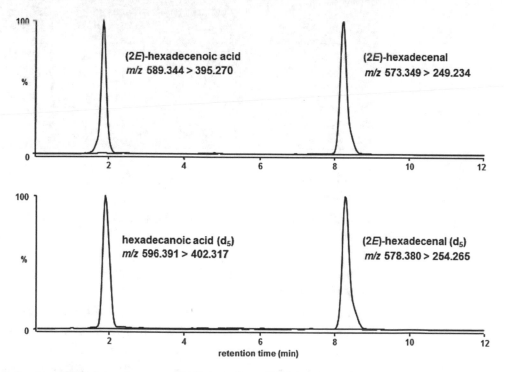

Fig. 3 Combined SRM chromatograms of DAIH derivatives of (2*E*)-hexadecenoic acid and (2*E*)-hexadecenal (upper chromatogram), and their stable-isotope labeled internal standards hexadecanoic acid (d_5) and (2*E*)-hexadecenal (d_5) (lower chromatogram). All compounds were analyzed in the same run with each peak set 100%

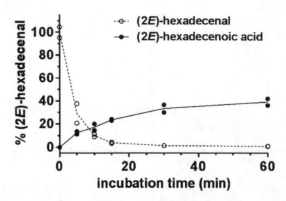

Fig. 4 Time course of the oxidation of (2*E*)-hexadecenal added to HepG2 cell lysates to (2*E*)-hexadecenoic acid after incubation at 37 °C. 100% correspond to 4 pmol of extracted and derivatized (2*E*)-hexadecenal injected into the LC-MS system. Quantification was performed using the isotope-dilution SRM approach. Data of two separate extractions (and subsequent derivatizations) from the same cell lysate are presented

Table 3
Settings of the MS/MS detector for simultaneous quantification of (2*E*)-hexadecenal and (2*E*)-hexadecenoic acid derivatives using isotope-dilution SRM

DAIH derivative of	Precursor ion [M–H]$^-$ (*m/z*)	Fragment ion [M–H]$^-$ (*m/z*)	Collision energy (V)
Fatty aldehydes			
(2*E*)-hexadecenal	573.3487	249.2336	24
(2*E*)-hexadecenal (d$_5$)	578.3800	254.2650	24
Fatty acids			
(2*E*)-hexadecenoic acid	589.3436	395.2704	30
hexadecanoic acid (d$_5$)	596.3906	402.3174	30

2. All aldehyde and fatty acid stock solutions are stable for at least 1 year when stored at −20 °C.

3. (2*E*)-hexadecenal (d$_5$) can also be solved in ethanol and stored at −20 °C.

4. DAIH is applied as solution close to the saturation level. If DAIH seems to be not completely dissolved, an extended sonification is recommended.

5. Prepare ethanolic pyridine solution under the fume hood.

6. The presented methodology for analysis of derivatized fatty aldehydes and fatty acids using an Agilent HPLC-ESI-QTOF instrument can be easily transferred to an equivalent HPLC-ESI-triple quadrupole machine like the Agilent 6490 Triple Quad LC/MS with iFunnel technology (from Waldbronn, Germany).

7. Always protect ears during sonification.

8. Recovery rate is higher the faster the Dole solution is added.

9. For derivatization of only $(2E)$-hexadecenal the addition of EDC hydrochloride can be omitted. In this case, dissolve the dried extract in 50 μL acetonitrile and add 182.5 μL DAIH solution and 17.5 μL 2 M HCl. Place the micro-inserts in new 1.5 mL screw-capped glass vials and cool the samples for 15 min at 4 °C prior to LC-MS analysis.

10. Incubation for 1 h at room temperature is recommended. While the derivatization reaction for fatty acids is completed after 15 min at 80 °C, it may happen that the aldehyde derivatization is not finished. Longer incubations at 80 °C come along with decreasing aldehyde-DAIH yields. But incubation at room temperature works well.

11. Derivatized samples are stable for up to 3 days when stored at 4 °C.

12. The limit of detection for $(2E)$-hexadecenal and $(2E)$-hexadecenoic acid was found to be 33 fmol and 69 fmol per injection using 4.5% fatty acid-free bovine serum albumin in phosphate buffered saline as a matrix. The limit of quantification was estimated to be 99 fmol per injection for $(2E)$-hexadecenal and 207 fmol per injection for $(2E)$-hexadecenoic acid. Method validation showed good intra- and interday precision [14].

References

1. Matsumoto S (2010) Molecular mechanisms underlying sex pheromone production in moths. Biosci Biotechnol Biochem 74(2)223–231

2. Palczewski K (2012) Chemistry and biology of vision. J Biol Chem 287(3)1612–1619

3. Goodenowe DB, Cook LL, Liu J, Lu Y, Jayasinghe DA, Ahiahonu PW, Heath D, Yamazaki Y, Flax J, Krenitsky KF, Sparks DL, Lerner A, Friedland RP, Kudo T, Kamino K, Morihara T, Takeda M, Wood PL (2007) Peripheral ethanolamine plasmalogen deficiency: a logical causative factor in Alzheimer's disease and dementia. J Lipid Res 48(11)2485–2498

4. Janfelt C, Wellner N, Leger PL, Kokesch-Himmelreich J, Hansen SH, Charriaut-Marlangue C, Hansen HS (2012) Visualization by mass spectrometry of 2-dimensional changes in rat brain lipids, including N-acylphosphatidylethanolamines, during neonatal brain ischemia. FASEB J 26(6)2667–2673

5. Senanayake VK, Jin W, Mochizuki A, Chitou B, Goodenowe DB (2015) Metabolic dysfunctions in multiple sclerosis: implications as to causation, early detection, and treatment, a case control study. BMC Neurol 15:154

6. Laurenzi VD, Rogers GR, Hamrock DJ, Marekov LN, Steinert PM, Compton JG, Markova N, Rizzo WB (1996) Sjögren-Larsson syndrome is caused by mutations in the fatty aldehyde dehydrogenase gene. Nat Genet 12(1)52–57

7. Nishida T, Vang VL, Yamazawa H, Yoshida R, Naka H, Tsuchida K, Ando T (2003) Synthesis and characterization of hexadecadienyl compounds with a conjugated diene system, sex pheromone of the persimmon fruit moth and related compounds. Biosci Biotechnol Biochem 67(4)822–829

8. Berdyshev EV (2011) Mass spectrometry of fatty aldehydes. Biochim Biophys Acta 1811(11)680–693

9. Ohshima T, Wada S, Koizumi C (1989) 1-O-alk-1′-enyl-2-acyl and 1-O-alkyl-2-acyl glycerophospholipids in white muscle of bonitoEuthynnus pelamis (Linnaeus) Lipids 24(5)363–370

10. Brahmbhatt V, Nold C, Albert C, Ford D (2008) Quantification of pentafluorobenzyl oxime derivatives of long chain aldehydes by GC–MS analysis. Lipids 43(3)275–280

11. Kölliker S, Oehme M, Dye C (1998) Structure elucidation of 2,4-dinitrophenylhydrazone derivatives of carbonyl compounds in ambient air by HPLC/MS and multiple MS/MS using atmospheric chemical ionization in the negative ion mode. Anal Chem 70(9)1979–1985

12. Zwiener C, Glauner T, Frimmel F (2002) Method optimization for the determination of carbonyl compounds in disinfected water by DNPH derivatization and LC–ESI–MS–MS. Anal Bioanal Chem 372(5–6)615–621

13. Lüth A, Neuber C, Kleuser B (2012) Novel methods for the quantification of (2E)-hexadecenal by liquid chromatography with detection by either ESI QTOF tandem mass spectrometry or fluorescence measurement. Anal Chim Acta 722:70–79

14. Neuber C, Schumacher F, Gulbins E, Kleuser B (2014) Method to simultaneously determine the sphingosine 1-phosphate breakdown product (2E)-hexadecenal and its fatty acid derivatives using isotope-dilution HPLC-electrospray ionization-quadrupole/time-of-flight mass spectrometry. Anal Chem 86(18)9065–9073

15. Pizzimenti S, Ciamporcero ES, Daga M, Pettazzoni P, Arcaro A, Cetrangolo G, Minelli R, Dianzani C, Lepore A, Gentile F, Barrera G (2013) Interaction of aldehydes derived from lipid peroxidation and membrane proteins. Front Physiol 4:242

16. Aldini G, Domingues MR, Spickett CM, Domingues P, Altomare A, Sánchez-Gómez FJ, Oeste CL, Pérez-Sala D (2015) Protein lipoxidation: detection strategies and challenges. Redox Biol 5:253–266

17. Blair IA (2008) DNA adducts with lipid peroxidation products. J Biol Chem 283(23) 15545–15549

18. Upadhyaya P, Kumar A, Byun H-S, Bittman R, Saba JD, Hecht SS (2012) The sphingolipid degradation product trans-2-hexadecenal forms adducts with DNA. Biochem Biophys Res Commun 424(1)18–21

19. Duerr MA, Aurora R, Ford DA (2015) Identification of glutathione adducts of α-chlorofatty aldehydes produced in activated neutrophils. J Lipid Res 56(5)1014–1024

20. Boon PJM, Marinho HS, Oosting R, Mulder GJ (1999) Glutathione conjugation of 4-hydroxy-trans-2,3-nonenal in the rat in vivo, the isolated perfused liver and erythrocytes. Toxicol Appl Pharmacol 159(3)214–223

21. Nakahara K, Ohkuni A, Kitamura T, Abe K, Naganuma T, Ohno Y, Zoeller Raphael A, Kihara A (2012) The Sjögren-larsson syndrome gene encodes a hexadecenal dehydrogenase of the sphingosine 1-phosphate degradation pathway. Mol Cell 46(4)461–471

22. Püttmann M, Krug H, von Ochsenstein E, Kattermann R (1993) Fast HPLC determination of serum free fatty acids in the picomole range. Clin Chem 39(5)825–832

23. Ciccimaro E, Blair IA (2010) Stable-isotope dilution LC–MS for quantitative biomarker analysis. Bioanalysis 2(2)311–341

Chapter 11

CE Analysis of Phospholipid Headgroups

Václav Matěj Bierhanzl, Martina Riesová, Gabriela Seydlová, and Radomír Čabala

Abstract

The capillary electrophoresis method has been developed for the analysis of phospholipid polar headgroups. The method consists in splitting the phospholipid samples by regioselective enzyme that cleaves the bond between polar headgroup and glycerol. The presented capillary electrophoresis method operates in basic pH where all the phosphoesters are negatively charged and their mobilities are different enough to allow their separation. Separation electrolyte needs to be defined by ionic strength to give repeatable results. Despite low absorbance of phosphoesters UV detection is more sensitive than the contactless conductometry one.

Key words Capillary electrophoresis, Phosphoesters, Phospholipid headgroups, Phospholipids

1   Introduction

Phosphoesters are a heterogenous group of esters of phosphoric acid which correspond to phospholipid classes (phosphoethanolamine, phosphoserine, etc.). They can be yielded from organic tissues by cleaving the phospholipids with phospholipase C, which splits the bond between phosphoester and the glycerol backbone. Phosphoesters are anions so they can be separated by the differences of their electrokinetic mobilities. The first attempts of electrophoretic analyses of phosphoesters are connected with amino acid research [1] where phosphoserine was determined [2]. Several other single phosphoester [3] were analyzed by different capillary electrophoresis mostly nitrogen containing phosphoesters [4]. Nevertheless, these methods were heterogenous, targeted only to few phosphoester analytes. Because phosphoesters almost do not absorb ultraviolet neither visible spectra [5], they were tested with conductometry detection, indirect UV detection [6] or with derivatization of fluorescent agent [7–9]. Finally, despite low absorbance of phosphoesters, the UV detection showed up to be more sensitive method than conductometry for a presented method for the analysis of this group [10].

Paul Wood (ed.), *Lipidomics*, Neuromethods, vol. 125,
DOI 10.1007/978-1-4939-6946-3_11, © Springer Science+Business Media LLC 2017

2   Materials

Prepare a Li/CHES buffer of 0.1 M CHES and 0.06 LiOH concentration. Use LiOH × H_2O as a stable Li^+ salt. Buffer solution should have about pH 10. Do not adjust the pH and use LiOH only (Fig. 1). Avoid NaOH or any other sodium ions. Avoid all hygroscopic materials also.

Prepare stock standards of phosphoesters (phosphoglycerol, phosphoethanolamine, phosphoserine) of 1 mg/mL. Use separation electrolyte (diluted 1:10 with distilled water) as a solvent. Use it as a diluent for making the calibration solutions.

Separation capillary should be of noncoated silica with 75 μm diameter, 59 cm length, and 50 cm effective length to obtain sufficient resolution and the runtime of cca 20 min when using separation voltage of 20 kV. In case of different voltage the length of capillary needs to be scaled to the same intensity of electric field.

Apparatus used for the analyses should have the autosampler, high voltage source that maintains the constant voltage in the range of 15 to 35 kV and the UV detector (or PDA/DAD detector). Combined detector of UV and CCD is optionable.

Prepare 1 M NaOH and 1 M HCl solutions for cleaning.

3   Method

Calibration row of three to five points should cover the interval of one or two orders of magnitude, typically 0.1 to 1 mg/mL.

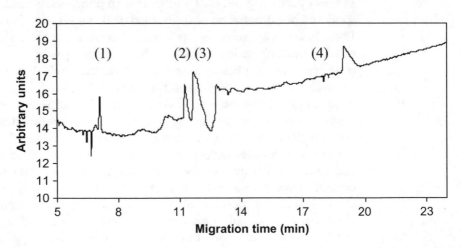

Fig. 1 Separation of phospholipid headgroups by capillary electrophoresis: (*1*) phosphoethanolamine, (*2*) phosphoserine, (*3*) phosphoglycerol, (*4*) phosphate. Separation electrolyte: 0.1 mol/L CHES, 0.06 mol/L LiOH; Separation voltage 20 kV; Capillary diameter 75 μm, 50 cm effective length, and 59 cm total length; Detection wavelength 205 nm

Separation capillary should be conditioned with separation electrolyte 10–15 min. It should be flushed with separation electrolyte 3 min before every run.

Separation capillary should be purged with 1 M NaOH and 1 M HCl after the run. Capillary has to be flushed (at least 3 min) with distilled water after every NaOH/HCl purging. Distilled water flushing of separation capillary is necessary to avoid crystallization of buffer or other salts.

Detection needs to be carried out at 205 nm wavelength (negative polarity).

Temperature of detection can be 20 °C or 25 °C (ambient temperature), but it is better to be defined to obtain better reproducibility of migration times.

Separation voltage should be set at 20 kV. Resolution between phosphoserine and phosphoglycerol must be 1.5 and more for the most concentrated point of calibration curve. Otherwise, intensity of electric field needs to be lowered.

Acknowledgment

This work was supported by SVV260317 and 13-18051P from the Czech Science Foundation.

References

1. Yan JX, Packer NH, Gooley AA, Williams KL (1998) Protein phosphorylation: technologies for the identification of phosphoamino acids. J Chromatogr A 808:23–41

2. Fadden P, Haystead TAJ (1995) Quantitative and selective fluorophore labeling of phosphoserine on peptides and proteins: characterization at the attomole level by capillary electrophoresis and laser-induced fluorescence. Anal Biochem 225:81–88

3. Robert F, Bert L, Parrot S, Denoroy L, Stoppini L, Renaud B (1998) J Chromatogr A 817:195–203

4. Wang K, Jiang D, Sims CE, Allbritton NL (2012) Coupling on-line brain microdialysis, precolumn derivatization and capillary electrophoresis for routine minute sampling of O-phosphoethanolamine and excitatory amino acids. J Chromatogr B 907:79–86

5. Liu X, Hu Y-Q, Ma L, Lu Y-T (2004) Determination of phosphoamino acids derivatized with 5-(4,6-dichloro-s-triazin-2-ylamino) fluorescein by micellar electrokinetic chromatography. J Chromatogr A 1049:237–242

6. Crowder MW, Numan A, Haddadian F, Weitzel MA, Danielson ND (1999) Capillary electrophoresis of phosphoamino acids with indirect photometric detection. Anal Chim Acta 384:127–133

7. Fu N-N, Wang H, Zhang H-S (2011) Rapid analysis of phosphoamino acids from phosvitin by near-infrared cyanine 1-(ε-succinimydyl-hexanoate)-1'-methyl-3,3,3',3'-tetramethyl-indocarbocyanine-5,5'-disulfonate potassium derivatization and polyacrylamide-coated CE with LIF detection. Electrophoresis 32:712–719

8. Zhang L-Y, Sun M-X (2007) Capillary electrophoresis of phosphorylated amino acids with fluorescence detection. J Chromatogr B 859:30–36

9. Liu X, Ma L, Lu Y-T (2004) Determination of phosphoamino acids by micellar electrokinetic capillary chromatography with laser induced fluorescence detection. Anal Chim Acta 512:297–304

10. Bierhanzl VM, Riesová M, Taraba L, Čabala R, Seydlová G (2015) Analysis of phosphate and phosphate containing headgroups enzymatically cleaved from phospholipids of Bacillus subtilis by capillary electrophoresis. Anal Bioanal Chem 407:7215–7220

Chapter 12

HPTLC-MALDI TOF MS Imaging Analysis of Phospholipids

Tatiana Kondakova, Nadine Merlet Machour, and Cécile Duclairoir Poc

Abstract

Phospholipids are major and essential functional components of all living cells playing fundamental roles in cellular metabolism and homeostasis. At molecular level, these cell compounds function as a barrier between the cell and its modifying environment and play an essential role in cell adaptation to environmental stressors. In the human body, phospholipids play a role of key metabolites in many pathways, either in health or in disease. However, because of the development of genomics and proteomics tools at the end of the twentieth century, these essential for all living cells molecules remain to be investigated in more detail. In this effort, we adopted a protocol, using MALDI-TOF MS Imaging coupled to HPTLC, to screen a large number of phospholipid classes in a short span of time. This method set the stage for future studies aimed at better defining the diversity and roles of phospholipids in all living cells.

Key words Lipidomics, Phospholipids, Mass spectrometry imaging, HPTLC MALDI TOF MSI

1 Introduction

Despite an excessive development of proteomics and genomics tools caused by the emergence of molecular biology techniques at the end of the twentieth century, there is also a growing interest in the analysis and identification of lipids [1]. All living cells, prokaryotic, as well as eukaryotic contain a myriad of lipids, whose roles are still poorly studied. Although a considerable diversity of lipid structures exists, most predominant are phospholipids (PLs) [2]. PLs are described as acylated derivatives of *sn*-glycerol-3-phosphate typically composed of two fatty acid chains, a glycerol unit, and a phosphate group linked to a polar head group [3].

PLs are the major components of all the biological membranes contributing to their biochemical and biophysical properties [2, 4, 5]. In mammalian cells, PLs play several roles, being often key metabolites in many pathways, either in health or in disease. The changes in membrane PLs' structure and composition are correlated to several diseases such as diabetes [6, 7], and Alzheimer's disease [8, 9]. The roles of PLs are especially remarkable in domains of neurology, neuro-oncology, and psychiatry. Recently, the PLs'

Paul Wood (ed.), *Lipidomics*, Neuromethods, vol. 125,
DOI 10.1007/978-1-4939-6946-3_12, © Springer Science+Business Media LLC 2017

contribution in depression and anxiety disorders [5], as well as in synaptic communication [10], was proposed. In this way, the roles of PLs in bacteria–host interactions should also be taken into account because of the possible correlation of the human micro-biota effect on the brain function [11], and multiple strategies used by bacteria to exploit eukaryotic PLs [2, 12, 13].

Altogether, these investigations encouraged development of lipidomics and metabolomics (i.e., analysis of endogenous bio-molecules, generally <2000 Da, in complex biological matrices) [14]. Mass spectrometric tools, including Matrix-Assisted Laser Desorption Ionization Time-Of-Flight Mass Spectrometry (MALDI TOF MS), are some of the most robust analytical methods that provide structural data for complex biological matrices [15, 16]. The use of MALDI TOF MS for the PLs analysis helps to overcome many problems related to the com-plexity and diversity of the biological extracts, and appears as one of the most convenient methods to realize a complete PL screening [17, 18].

Here, we describe an analytical method for the PLs' study based on the direct coupling High Performance Thin Layer Chromatography (HPTLC) to MALDI TOF MS Imaging. This technique, adapted for the analysis of bacterial lipids, appears as a universal [15, 19], rapid, and globally cheap method for the PLs screening in all living cells. Together with another lipidomic tech-niques, which can be found in this volume of the Series, these protocols should be useful to explore the PL composition, local-ization, and role(s) in both eukaryotic and prokaryotic cells and, thus, better assess the PLs functions either in health or in disease, elucidating, among others, the role(s) of PLs in host–parasite and/or host–symbiont interactions.

2   Materials

2.1   Phospholipid Extraction

1. Extraction solution A: $CHCl_3/CH_3OH$ (1/2, v/v). Mix solution for ~1 min using vortexer to obtain the homogenous mixture (*see* **Note 1**).

2. Chloroform (*see* **Note 1**).

3. Water (*see* **Note 2**).

4. Centrifuge.

2.2   High-Performance Thin-Layer Chromatography (HPTLC)

1. Elution system: $CHCl_3/CH_3-CH_2OH/H_2O/N(CH_2-CH_3)_3$ (35/35/7/35, v/v/v/v) (*see* **Notes 1** and **3**).

2. HPTLC silica gel 60 plates F_{254} (75 × 50 mm, on aluminum backs) obtained from Merck (Darmstadt, Germany).

3. TLC chambers (e.g., TLC TANK 80 × 120 mm from Fisher Scientific SAS).

4. Primuline dye: 0.05% solution in CH_3-CO-CH_3/H_2O, (8/2, v/v) (*see* **Note 1**).

2.3 HPTLC-MALDI TOF Analysis

1. DHB matrix solution: 200 g/L 2,5-dihydroxybenzoic acid (DHB) in C_2H_3N/0.1% trifluoroacetic acid (TFA), (90/10, v/v) [19] (*see* **Notes 1** and **4**). 10 mL volume is recommended for coating of one HPTLC plate.

2. TLC MALDI target from Bruker Daltonics (Bremen, Germany).

3. Time-of-flight mass spectrometer (e.g., Autoflex III from Bruker Daltonics, Bremen, Germany).

4. TLC MALDI software (e.g., v. 1.1.7.0 from Bruker Daltonics, Bremen, Germany).

5. Peptide Calibration Standard II from Bruker Daltonics (Bremen, Germany).

6. MALDI Imaging software (e.g., FlexImaging software v. 2.1., Bruker Daltonics, Bremen, Germany).

3 Methods

3.1 Phospholipid Extraction

The quickest and simplest method to extract PLs from living cells has been described by Bligh and Dyer [20] (Fig. 1). This procedure could be applied to all biological materials (i.e., fish muscle [20], mitochondria [21], protozoa [22], and bacteria [23, 24]) and can be carried out in approximately 10 min. Briefly, a purified PL extract is obtained by homogenization of samples with a mixture of chloroform and methanol ($CHCl_3$/CH_3OH) in such proportions that a miscible system is formed with the water in sample. Additional dilution with chloroform and water separates the homogenate into two layers, among which the lower chloroform layer contains PLs.

1. To a 1 g sample (*see* **Notes 5** and **6**) add 3 mL of the extraction solution A and vortex during 3 min.

2. Add 1 mL of chloroform and vortex during 1 min.

3. Add 1 mL of water (*see* **Note 2**) and vortex during 3–5 min.

4. Centrifuge the mixture 10 min at 3000 × g to allow the formation of the two phases (i.e., upper aqueous phase and lower chloroform phase).

5. Record the lower chloroform layer in another glass tube (*see* **Note 7**) using a Pasteur pipette.

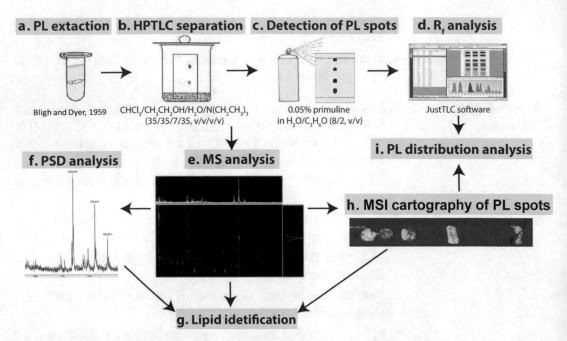

Fig. 1 Scheme illustrating the general HPTLC-MALDI TOF MSI method of phospholipid analysis. After extraction (**a**), phospholipids (PL) are separated on HPTLC plate (**b**) and detected using primuline spray (**c**), conducting to calculation of PL spots' retention factors (R_f) (**d**). MS analysis (**e**) of PL spots is directly performed on HPTLC plate in TLC MALDI software allowing the direct PDS analyses of each PL m/z (**f**) and PL identification (**g**). MALDI MS Imaging cartography of PL spots (**h**) is performed to study the PL distribution on HPTLC plate (**i**) according to their m/z, confirming PL identification

6. Evaporate chloroform under N_2 stream (*see* **Note 8**).

7. Measure the total lipid weight.

8. Store the extracted lipids at −20°C under N_2 atmosphere (*see* **Note 9**).

3.2 High-Performance Thin-Layer Chromatography (HPTLC)

High-Performance Thin-Layer Chromatography (HPTLC) is the simplest and well-known technique used in our work to separate individual PL species before MS and Post Source Decay (PSD) analyses (Fig. 1). We routinely use HPTLC silica gel 60 plates F_{254}, which allow PLs' separation according to their polar head groups (HGs) and are adapted to the future MALDI analyses. Our standard procedure is described below:

1. Wash HPTLC plate using elution system. For this, put the plate in a TLC chamber containing by a few mL of elution system and do the solvent system migration up to the top of the HPTLC plate (*see* **Note 1**).

2. Activate the HPTLC plate at 110 °C during 2 h (*see* **Note 10**).

3. To previously extracted PLs add chloroform in such a quantity to obtain the total PLs concentration ~1.5 mg/mL.

4. Spot 100 μL of lipid extract on the HPTLC plate by spotting of 2 μL 50 times (*see* **Note 11**).

5. Separate PL spots using $CHCl_3/CH_3\text{-}CH_2OH/H_2O/N(CH_2\text{-}CH_3)_3$ (35/35/7/35, v/v/v/v) as running separation solution (*see* **Note 1**) (Fig. 1b).

6. Remove HPTLC plate from the TLC chamber and dry the plate for 15 min using the chemical hood.

7. Spray primuline dye on HPTLC plate and dry for 15 min under chemical hood (Fig. 1c).

8. Visualize PL spots by UV fluorescence at 365 nm (*see* **Note 12**) (Figs. 3a and 4a).

9. Calculate the retention factors (R_f) of PL spots. For R_f calculation we usually use the Sweday JustTLC software (v. 4.0.3, Lund, Sweden) (Fig. 1d).

3.3 HPTLC-MALDI TOF Analysis

After separation on HPTLC plate, PLs are directly analyzed by MALDI TOF MS using TLC MALDI target (Fig. 2a). The hyphenated HPTLC-MALDI approach requires the uniform coverage of the TLC lanes with MALDI matrix and allows direct MS and PSD analyses of PLs previously separated on HPTLC plate (Fig. 1e and f). The mass spectrometric information is subsequently read out automatically using TLC software, allowing the screening of one sample in about 5 min. MALDI plate movement along the chromatographic lanes is driven by the standard movement mechanics of the MALDI ion source without additional robotic requirements. The quick analysis of PLs' MS spectra is also possible thanks to an interactive MALDI Chromatogram. MALDI MS

A TLC MALDI target

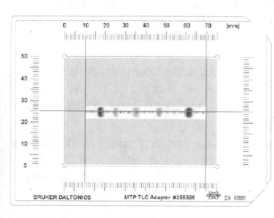

B TLC MALDI software

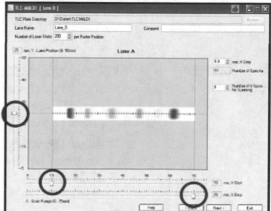

Fig. 2 HPTLC-MALDI coupling. (**a**) Scheme of TLC-MALDI adapter target. Orange and purple lines indicate the position of chromatographic line. (**b**) TLC-MALDI setup dialog

Imaging (Fig. 1h), designed for MS analysis of tissue, is also a good method for mapping individual PL species distribution in HPTLC plates, completing analysis.

3.3.1 HPTLC-MALDI TOF Coupling

1. Fill glass Petri dish with DHB matrix (200 mg/mL in $C_2H_3N/0.1\%$ TFA (90/10, v/v)) (*see* **Note 4**).

2. Submerge the plate in the reservoir and remove it immediately (typical immersion time: 1 s).

3. Lay the plate on a clean surface and dry it in two steps.
   First step: Dry the plate with a very gentle airstream (e.g., use airstream close to a fume hood window) around 2 min until the surface becomes matt.
   Second step: Dry the HPTLC plate with a hair-dryer in a stream of cold air vertically from above for about 90 s.

4. Repeat matrix application and drying procedure. This time hold the plate on the opposite edge and blow in the second step for 4 min (*see* **Notes 13** and **14**).

5. Insert the HPTLC plate into the dedicated adapter target (e.g., TLC MALDI target from Bruker Daltonics, Bremen, Germany) (Fig. 2a).

6. Put the HPTLC plate with the target in the desiccator for a few minutes (*see* **Note 15**).

3.3.2 Calibration

Automatic MALDI TOF calibration is performed in two steps.

1. External calibration of the apparatus is performed daily in FlexControl software using the DHB matrix peaks and the Peptide Calibration Standard II (ref 822570, Bruker Daltonics, Bremen, Germany) covering a mean mass range between 200 and 3500 Da (*see* **Note 16**).

2. Internal calibration was performed to calibrate the acquired spectra in TLC MALDI software and made using two DHB matrix peaks at 171.2 m/z and 273.0 m/z.

3.3.3 MALDI TOF Analysis

1. Using TLC MALDI software, delimit the position of PL spots and define the number of laser shoots per point (we routinely use 200 laser shoots per point). Specify the distance between points (1 mm) (Fig. 2b).

2. Adjust the voltage characteristics and laser power in FlexControl software (*see* **Note 17**). The extraction voltages are 19.50 and 17.30 kV. The reflector voltages are 21 and 9.40 kV. All spectra are acquired in positive reflector mode using delayed extraction.

3. Adjust the laser power about 30% above threshold to have a good signal-to-noise ratio. A major strength is needed to desorb lipids from silica [25].

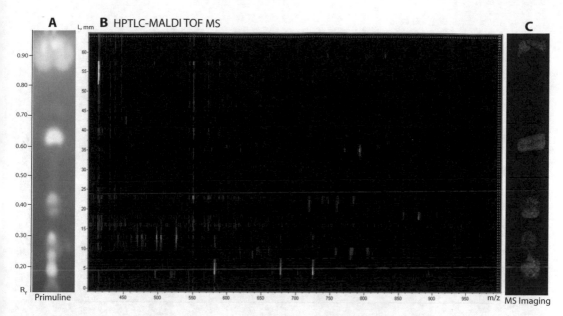

Fig. 3 HPTLC-MALDI TOF MSI analysis. (**a**) Primuline staining of separated lipids on a HPTLC plate. (**b**) HPTLC MS readout. X- and y-axis correspond to the m/z values and the lipid spots respectively. The MS peaks' intensity is presented by red-to-yellow scale. (**c**) HPTLC-MALDI MSI analysis of identical HPTLC plate. According MS and PSD analyses, PL spots are reconstructed by color-labeling according to their m/z

4. Start an automatic spectra acquisition using TLC MALDI software. Obtained by this technique MALDI chromatogram enables the rapid (about 5 min) MS analysis of each sample (Fig. 3c). Y-axis corresponds to the chromatographic line (PL spots) (Fig. 4d) and X-axis presents MS peak intensity (increasing m/z) (Fig. 4c). To eliminate most of the DHB matrix peaks, only m/z from 500 to 2000 were studied.

5. Select and analyze the m/z of PL spots according to the peaks' intensity showing on the right of TLC MALDI software and the MS spectra of each PL spot showing at the top of the window (Fig. 4c and d).

6. In order to identify the PLs, PSD spectra are acquired as previously described [26]. The precursor ions are isolated using a time ion selector. The fragment ions are refocused onto the detector by stepping the voltage applied to the reflectron in appropriate increments. It is done automatically by using the "FAST" ("fragment analysis and structural TOF") subroutine of the FlexAnalysis software (*see* **Note 17**). The obtained MS and PSD spectra are then identified using the LIPID MAPS database.

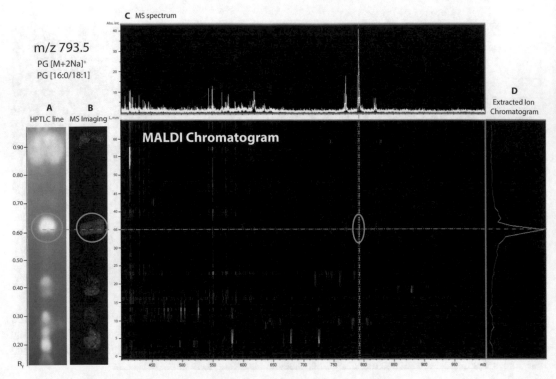

Fig. 4 HPTLC-MALDI TOF MS analysis of phosphatidylglycerol (PG). PG [16:0/18:1] with observed m/z 793.5, corresponding to PG [M + 2Na]+, is shown as an example. (**a**) Primuline straining of separated lipids on a HPTLC plate. R_f of PG spot = 0.62 ± 0.03 (*blue circle*). (**b**) HPTLC-MALDI MSI analysis of identical HPTLC plate, showing the PG spot (m/z 793.5, rose spot in *blue circle*). (**c**) MS spectrum of R_f = 0.62 allows identification of of PG at m/z 793.5 (*blue dashed line*). (**d**) Extracted ion chromatogram showing the intensity of PG peak at m/z 793.5. The (**c**) and (**d**) data are combined in the MALDI Chromatogram

3.3.4 HPTLC-MALDI TOF Imaging

1. Using FlexImaging software define a polygon measurement region as the HPTLC line of each sample from the PL extract depot to the solvent front.

2. Acquisition method settings: we recommend averaging 500 single laser shots for each mass spectrum (raster width 200 μm). Multiple additions of single position acquisition run (every 40 shots) are employed to obtain a minimal spectrum intensity scale of 10^4 ion counts. The obtained spectra are automatically smoothed and baselined to limit the background noise. The voltage characteristics and laser power are stetted as previously described (*see* Sect. 3.3.3).

3. Start spectra acquisition.

4. After the end of spectra acquisition, edit the mass filters for PLs according to their m/z identified previously (*see* Sect. 3.3.3) (Figs. 3b and 4b).

4   Notes

1. All chemicals are of the highest commercially available purity and used without any further purification. Solutions are best prepared in advance and conserved at room temperature.

2. We recommend using ultrapure HPLC grade water.

3. Elution system should be colorless and transparent.

4. DHB is selected as a MALDI matrix because of high signal strength and the absence of matrix adduct peaks. PSD analysis with DHB matrix allows the cleavage of phosphate-glycerol bond and induces the loss of the hydrophilic head group, i.e., $[M-HG + H]^+$, characteristic of each lipid class. This loss is caused by a positive charge localized on the phosphate group [27].

5. For phospholipid extraction, we routinely use lyophilized samples. However, Blight and Dyer's method is also adapted for samples, containing $80 \pm 1\%$ water and 1% lipids [20].

6. For tissues' study, we highly recommend homogenizing the sample before lipid extraction.

7. A small volume of the chloroform layer must be let with aqueous phase. We usually prefer to lose a few microliters of chloroform phase rather than risking contaminating PL stocks.

8. Chloroform evaporation is necessary to unsure (1) the same volume in each lipid sample and (2) the correct condition for PLs' storage. Diluted in chloroform PLs are easer oxidized during storage than dried ones.

9. The usual PLs' storage time is about 6 months.

10. "Activation" is evaporating the water in the silica. The silica gel is hygroscopic, so it adsorbs water vapor from the air and becomes hydrated. This hydration in turn, can alter chromatography and interaction of solvent system with sample.

11. All analyses are done in triplicate. We routinely spot three PL extracts on the same HPTLC plate analyzing three replicates at the same time.

12. Primuline binds noncovalently to the apolar fatty acyl residues of lipids and does not affect a subsequent MS analysis [28].

13. With two dips you get a mass occupancy of about 5 mg/cm². To date, good results have been achieved by dip-coating twice. Every additional coating requires longer drying time and more repetitions can cause bleeding between the separated bands.

14. After the matrix coating the HPTLC plate should look smooth and flat. If the surface looks rough and blistered either the

organic solvent concentration in the matrix solution was too low or the drying process was not quick enough.

15. Desiccation of the HPTLC plate is likely to quickly create a vacuum in mass spectrometer.

16. We usually do the external calibration just before starting HPTLC-MALDI analysis. Thus, 1 µL of the Peptide Calibration Standard II is directly spotted on the TLC MALDI target and coated by the 1 µL DHB matrix.

17. The system utilizes an Autoflex III mass spectrometer equipped with a laser OptibeamTM Nd/YAG (355 nm) with 200-Hz tripled-frequency (Bruker Daltonics, Bremen, Germany).

Acknowledgments

TK is a recipient of a PhD grant from the GRR SeSa (Sanitary Safety Research Network) financed by the Regional Council of Haute-Normandie (France). This study was supported by the Conseil Général de l'Eure and Grand Evreux Agglomeration.

References

1. Gidden J, Denson J, Liyanage R, Ivey DM, Lay JO (2009) Lipid compositions in *Escherichia coli* and Bacillus subtilis during growth as determined by MALDI-TOF and TOF/TOF mass spectrometry. Int J Mass Spectrom 283:178–184. doi:10.1016/j.ijms.2009.03.005

2. Kondakova T, D'Heygère F, Feuilloley MJ, Orange N, Heipieper HJ, Duclairoir Poc C (2015) Glycerophospholipid synthesis and functions in Pseudomonas. Chem Phys Lipids 190:27–42. doi:10.1016/j.chemphyslip.2015.06.006

3. Zhang Y-M, Rock CO (2008) Membrane lipid homeostasis in bacteria. Nat Rev Microbiol 6:222–233. doi:10.1038/nrmicro1839

4. O'Brien JS, Sampson EL (1965) Lipid composition of the normal human brain: gray matter, white matter, and myelin. J Lipid Res 6:537–544

5. Müller CP, Reichel M, Mühle C, Rhein C, Gulbins E, Kornhuber J (2015) Brain membrane lipids in major depression and anxiety disorders. Biochim Biophys Acta 1851:1052–1065. doi:10.1016/j.bbalip.2014.12.014

6. Naudí A, Jové M, Ayala V, Cabré R, Portero-Otín M, Pamplona R (2013) Non-enzymatic modification of aminophospholipids by carbonyl-amine reactions. Int J Mol Sci 14:3285–3313. doi:10.3390/ijms14023285

7. Solis-Calero C, Ortega-Castro J, Frau J, Munoz F (2015) Nonenzymatic reactions above phospholipid surfaces of biological membranes: reactivity of phospholipids and their oxidation derivatives. Oxid Med Cell Longev 2015(2015):e319505. doi:10.1155/2015/319505, 10.1155/2015/319505

8. Astarita G, Jung K-M, Berchtold NC, Nguyen VQ, Gillen DL, Head E et al (2010) Deficient liver biosynthesis of docosahexaenoic acid correlates with cognitive impairment in Alzheimer's disease. PLoS One 5(9):e12538. doi:10.1371/journal.pone.0012538

9. Wood PL (2012) Lipidomics of Alzheimer's disease: current status. Alzheimer's Res Ther 4:5. doi:10.1186/alzrt103

10. García-Morales V, Montero F, González-Forero D, Rodríguez-Bey G, Gómez-Pérez L, Medialdea-Wandossell MJ et al (2015) Membrane-derived phospholipids control synaptic neurotransmission and plasticity. PLoS Biol 13:e1002153. doi:10.1371/journal.pbio.1002153

11. Foster JA, Lyte M, Meyer E, Cryan JF (2016) Gut microbiota and brain function: an evolving field in neuroscience. Int J Neuropsychopharmacol 19(5). doi:10.1093/ijnp/pyv114

12. van der Meer-Janssen YPM, van Galen J, Batenburg JJ, Helms JB (2010) Lipids in host–

pathogen interactions: pathogens exploit the complexity of the host cell lipidome. Prog Lipid Res 49:1–26. doi:10.1016/j.plipres.2009.07.003

13. Vromman F, Subtil A (2014) Exploitation of host lipids by bacteria. Curr Opin Microbiol 17:38–45. doi:10.1016/j.mib.2013.11.003

14. Wood PL (2014) Mass spectrometry strategies for clinical metabolomics and lipidomics in psychiatry, neurology, and neuro-oncology. Neuropsychopharmacology 39:24–33. doi:10.1038/npp.2013.167

15. Schiller J, Süß R, Fuchs B, Muller M, Zschornig O, Arnold K (2007) MALDI-TOF MS in lipidomics. Front Biosci 12:2568–2579

16. Fuchs B, Schiller J, Süss R, Schürenberg M, Suckau D (2007) A direct and simple method of coupling matrix-assisted laser desorption and ionization time-of-flight mass spectrometry (MALDI-TOF MS) to thin-layer chromatography (TLC) for the analysis of phospholipids from egg yolk. Anal Bioanal Chem 389:827–834. doi:10.1007/s00216-007-1488-4

17. Fuchs B, Süß R, Nimptsch A, Schiller J (2009) MALDI-TOF-MS directly combined with TLC: a review of the current state. Chromatographia 69:95–105. doi:10.1365/s10337-008-0661-z

18. Fuchs B, Süß R, Schiller J (2010) An update of MALDI-TOF mass spectrometry in lipid research. Prog Lipid Res 49:450–475. doi:10.1016/j.plipres.2010.07.001

19. Fuchs B, Schiller J, Süss R, Zscharnack M, Bader A, Müller P et al (2008) Analysis of stem cell lipids by offline HPTLC-MALDI-TOF MS. Anal Bioanal Chem 392:849–860. doi:10.1007/s00216-008-2301-8

20. Bligh EG, Dyer WJ (1959) A rapid method of total lipid extraction and purification. Biochem Cell Biol 37:911–917. doi:10.1139/o59-099

21. Angelini R, Vitale R, Patil VA, Cocco T, Ludwig B, Greenberg ML et al (2012) Lipidomics of intact mitochondria by MALDI-TOF/MS. J Lipid Res. doi:10.1194/jlr.D026203

22. Palusinska-Szysz M, Kania M, Turska-Szewczuk A, Danikiewicz W, Russa R, Fuchs B (2014) identification of unusual phospholipid fatty Acyl compositions of *Acanthamoeba castellanii*. PLoS One 9:e101243. doi:10.1371/journal.pone.0101243

23. Lopalco P, Angelini R, Lobasso S, Köcher S, Thompson M, Müller V et al (2013) Adjusting membrane lipids under salt stress: the case of the moderate halophilic organism Halobacillus halophilus. Environ Microbiol 15:1078–1087. doi:10.1111/j.1462-2920.2012.02870.x

24. Kondakova T, Merlet-Machour N, Chapelle M, Preterre D, Dionnet F, Feuilloley M et al (2014) A new study of the bacterial lipidome: HPTLC-MALDI-TOF imaging enlightening the presence of phosphatidylcholine in airborne *Pseudomonas fluorescens* MFAF76a. Res Microbiol. doi:10.1016/j.resmic.2014.11.003

25. Lobasso S, Lopalco P, Angelini R, Vitale R, Huber H, Müller V et al (2012) Coupled TLC and MALDI-TOF/MS analyses of the lipid extract of the hyperthermophilic archaeon Pyrococcus furiosus. Archaea 2012:1–10. doi:10.1155/2012/957852

26. Fuchs B, Schober C, Richter G, Süß R, Schiller J (2007) MALDI-TOF MS of phosphatidylethanolamines: different adducts cause different post source decay (PSD) fragment ion spectra. J Biochem Biophys Methods 70:689–692. doi:10.1016/j.jbbm.2007.03.001

27. Harvey DJ (1995) Matrix-assisted laser desorption/ionization mass spectrometry of phospholipids. J Mass Spectrom 30:1333–1346. doi:10.1002/jms.1190300918

28. Richter G, Schober C, Süß R, Fuchs B, Müller M, Schiller J (2008) The reaction between phosphatidylethanolamines and HOCl investigated by TLC: Fading of the dye primuline is induced by dichloramines. J Chromatogr B 867:233–237. doi:10.1016/j.jchromb.2008.04.010

Chapter 13

Global UHPLC/HRMS Lipidomics Workflow for the Analysis of Lymphocyte Suspension Cultures

Candice Z. Ulmer, Richard A. Yost, and Timothy J. Garrett

Abstract

Untargeted cellular lipidomics workflows should include key steps on sample handing, cell rinsing, lipid extraction, reconstitution solvents, and parameters for UHPLC/HRMS analysis. The challenge with implementing a comprehensive lipidomics workflow is that many of these individualized protocols have not been standardized or optimized within the metabolomics/lipidomics field. The UHPLC analysis of lipid extracts provides a platform to analyze many lipid species, eliminating the spectral overlap and deconvolution issues present in direct infusion experiments. The combination of UHPLC with high-resolution mass spectrometry (HRMS) enhances the ability to resolve, profile, and identify multiple lipid species that encompass a large dynamic range (Bird et al Anal Chem 83(3):940–949, 2010; Wang et al Rapid Commun Mass Spectrom 28(20):2201, 2014). This work describes a global lipidomics protocol aimed at detecting polar and nonpolar lipid species from lymphocyte suspension cultures by UHPLC/HRMS analysis in positive and negative ionization modes.

Key words Lipidomics, Lymphocyte suspension cells, Sample preparation, Folch lipid extraction, UHPLC/HRMS

1 Introduction

Lipidomics, a rapidly advancing subset of metabolomics, is the study of lipid metabolism and function within a biological system [1–4]. Lipidomic studies provide insight into elucidating drug mechanisms [5–8], better understanding disease etiology [4, 9], and developing biomarkers for a biological system [10–13]. The general aim of untargeted lipidomic studies is to extract and analyze all lipid species specific to a biological sample in an unbiased manner, regardless of the chemical diversity.

Untargeted workflows for lymphocyte suspension cultures should include: sample handling (e.g., collection, enzymatic quenching, and storage conditions), cell washing, pre-data acquisition normalization, metabolite and/or lipid extraction, reconstitution, and parameters for UHPLC/HRMS data acquisition.

Paul Wood (ed.), *Lipidomics*, Neuromethods, vol. 125,
DOI 10.1007/978-1-4939-6946-3_13, © Springer Science+Business Media LLC 2017

Unfortunately, sample preparation is a major source of variation in untargeted lipidomics studies. Standardized sample preparation protocols for untargeted cellular lipidomics studies have not been optimized as these procedures should be organism-dependent and cell-structure dependent (adherent or suspension) [14].

The most common methods for cell quenching coupled to MS analysis include a buffer-free aqueous solution containing an organic solvent or a buffered aqueous solution containing a volatile salt [15]. Either quenching solution can be maintained at an extreme cold or hot temperature to halt enzymatic activity, decrease the uptake of components in the extracellular medium, and slow down the secretion of metabolites/lipids. Additionally, quenching can be achieved by centrifugation at a low temperature between 1 and 5 °C. In the latter case, the quenching step can be combined with the cell washing procedure during centrifugation. The cell washing buffer solution should maintain the integrity of the cellular membrane, have electrospray ionization (ESI) compatibility, and preserve the metabolite recovery/yield. Unfortunately, cell rinsing solutions for the mass spectrometric analysis of suspension-cultured mammalian cells are not consistent and many are not optimal for ESI-LC-MS. Ammoniated cell washing buffers should be used during the cell rinsing step, as shown in Fig. 1, for their volatility, compatibility with ESI, and limited intracellular metabolite leakage.

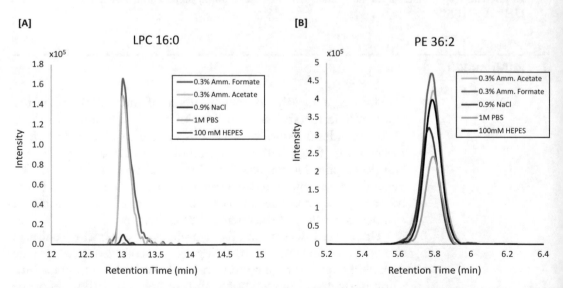

Fig. 1 Extracted ion chromatogram of endogenous lipids, (**a**) lysophosphatidylcholine 16:0 and (**b**) phosphatidylethanolamine 36:2 for each rinsing solution (0.3% ammonium formate, 0.3% ammonium acetate, 0.9% NaCl, 1 M PBS, 100 mM HEPES). More polar lipids were substantially affected by the presence of HEPES and nonvolatile salts in the cell rinsing solvent. Cell rinsing steps should be performed using an ammoniated cell washing buffer solution

Common choices for pre or post-data acquisition normalization of cell extracts include total cell count, protein concentration, or DNA concentration. Research has shown each method individually or combined (cell count normalization pre-data acquisition and protein/DNA normalization post-data acquisition) to provide great linearity in normalizing metabolomic datasets [16].

Many lipid liquid-liquid extraction solvent systems have been evaluated for extraction efficiency. No single extraction and/or separation method for untargeted lipidomics studies exists because lipids encompass a broad physical/chemical spectrum, ranging in properties such as acidity, stability, and polarity [16]. Nevertheless, the Folch method is considered the gold-standard and for the purpose of this work, is the most suitable to untargeted lipidomics studies of lymphocytes [14, 17]. This work presents an optimized lipidomics sample handling and sample preparation methodology for the untargeted UHPLC/HRMS analysis of lipids in lymphocyte suspension cells.

2   Materials

Prepare all solutions using LC/MS grade solvents. Store cell washing buffer in a refrigerator and cell pellets in temperatures at −80 °C or below. Triacylglyceride lipid standards (TG 15:0/15:0/15:0 and TG 17:0/17:0/17:0) were purchased from Sigma-Aldrich (St. Louis, MO). Exogenous lysophosphatidylcholine (LPC 17:0 and LPC 19:0), phosphatidylcholine (PC 17:0/17:0 and PC 19:0/19:0), phosphatidylethanolamine (PE 15:0/15:0 and PE 17:0/17:0), phosphatidylserine (PS 14:0/14:0 and PS 17:0/17:0), and phosphatidylglycerol (PG 14:0/14:0 and PG 17:0/17:0) lipid standards were purchased from Avanti Polar Lipids (Alabaster, Al). All lipid standards were dissolved prior to the analysis in 1:2 (v/v) chloroform:methanol ($CHCl_3$:MeOH) to make a 1000 ppm stock solution and a working 100 ppm standard mix was then prepared by diluting the stock solution with the same solvent mixture. All analytical grade solvents (formic acid, chloroform, and methanol) were purchased from Fisher-Scientific (Fairlawn, NJ). All mobile phase solvents were Fisher Optima LC/MS grade (acetonitrile, isopropanol, and water).

2.1 Suspension Cell Handling (Media Removal and Storage)

1. Lymphocyte suspension cells: 1×10^6 cells (*see* **Note 1**).

2. Cell rinsing solution: deionized water with 0.3% ammonium formate (*see* **Note 2**).

2.2 Lipid Extraction

1. Lipid standard mix stock solution (100 ppm in chloroform:methanol 1:2 (v/v)): LPC 17:0 and LPC 19:0, PC 17:0/17:0 and PC 19:0/19:0, PE 15:0/15:0 and PE

17:0/17:0, PS 14:0/14:0 and PS 17:0/17:0, and PG 14:0/14:0 and PG 17:0/17:0, TG 15:0/15:0/15:0 and TG 17:0/17:0/17:0

2. Extraction solvents: methanol (2 mL), chloroform (4 mL), water (1.5 mL). Each solvent is added separately during the extraction process.

3. Re-extraction solvent: chloroform/methanol (2:1, v/v). To prepare a stock solution, mix 5 mL of methanol and 10 mL of deionized water (*see* **Note 3**).

2.3 UHPLC/MS Analysis of Lipid Extracts

1. C18 column (75 × 2.1 mm, 1.9 μm) (*see* **Note 4**).

2. Reconstitution solvent: 100% isopropanol.

3. Solvent A: acetonitrile:water (60:40, v/v) with 10 mM ammonium formate and 0.1% formic acid.

4. Solvent B: isopropanol:acetonitrile:water (90:8:2, v/v) with 10 mM ammonium formate and 0.1% formic acid (*see* **Note 5**).

5. LC/MS analysis was performed with a Dionex Ultimate 3000 UHPLC system coupled to a Q Exactive™ hybrid quadrupole-orbitrap mass spectrometer operated in HESI-positive and negative ion mode.

3   Methods

3.1 Suspension Cell Handling (Media Removal and Storage)

The purpose of the cell washing step is to remove the culture media or extracellular components present in the cell matrix. The presence of these components (inorganic salts, anions, etc.) may mask the instrument signals for lipids and potentially cause ion suppression in the electrospray ionization process [14, 15].

1. Centrifuge the cell suspension from the culture dish/flask in a 15 mL conical tube at 311 × g for 5 min at 4 °C to pellet the lymphocyte cells. Discard the supernatant (*see* **Note 6**).

2. Wash cell pellet by adding 1 mL of the ice-cold cell rinsing solution directly to the pellet.

3. Centrifuge at 311 × g for 5 min at 4 °C. Discard the supernatant.

4. Repeat **steps 2** and **3** two more times (*see* **Note 7**).

5. Obtain a 5 μL aliquot for protein and/or DNA normalization measurements (*see* **Note 8**).

6. Store the lymphocyte cell pellet at −80 °C or lower or perform the lipid extraction for LC/MS analysis (*see* **Note 9**).

3.2 Lipid Extraction

Carry out the Folch lipid extraction procedure for lymphocyte cells with the samples and organic solvents kept on ice. Avoid having samples exposed to room temperature for more than 5 min.

Perform the Folch lipid extraction in the same 15 mL conical tube in which the cells were collected.

1. Spike in 15 μL of the 100 ppm lipid standard mix into cell pellet (*see* **Note 10**).

2. Add ice-cold methanol (2 mL) and chloroform (1 mL) directly to the cell pellet (*see* **Note 11**).

3. Vortex and incubate the sample on ice for 30 min (*see* **Note 12**).

4. Add ice-cold water (1.5 mL), vortex, and incubate the sample on ice for 10 min (*see* **Note 13**).

5. Centrifuge cells at 311 × *g* for 5 min at 4 °C to clearly separate the aqueous and organic layers (*see* **Note 14**).

6. Transfer the lower phase (organic layer) to a separate 5 or 15 mL conical tube using a glass pipette.

7. Add the re-extraction solvent (1 mL) to the aqueous layer, vortex, and centrifuge at 311 × *g* for 5 min at 4 °C (*see* **Note 15**).

8. Combine the organic layers.

9. Dry down the organic layer under nitrogen at 30 °C. For these studies, a MultiVap 118 nitrogen dryer (Organomation Associates, Inc.) was used (*see* **Note 16**).

3.3 UHPLC/HRMS Analysis of Lipid Extracts

Carry out all procedures at room temperature unless otherwise specified.

1. Reconstitute the dried lipid extract from the organic layer with 50 μL of isopropanol, vortex and centrifuge to mix.

2. Transfer lipid extracts to LC vials with 200 μL conical glass inserts.

3. Separation was performed using a Supelco Analytical Titan C18 column (2.1 × 75 mm, 1.9 μm) equilibrated at 30 °C with Solvents A and B as mobile phases (*see* **Note 16**).

4. The gradient used included 32% B at 0 min, 40% B at 1 min, a hold at 40% B until 1.5 min, 45% B at 4 min, 50% B at 5 min, 60% B at 8 min, 70% B at 11 min, and 80% B at 14 min at a flow rate of 0.5 mL/min (*see* **Note 17**).

5. Autosampler was maintained at 5 °C. The injection volume was 5 μL.

6. The following MS conditions employed for positive (negative) ion mode are included: spray voltage, 3.5 (3.5) kV; sheath gas, 30 (25) arbitrary units; sweep gas, 1 (0) arbitrary units; auxiliary nitrogen pressure, 5 [15] arbitrary units; capillary temperature, 300 (250)°C; HESI auxiliary gas heater temperature 350 (350)°C, and S-lens RF, 35 (35) arbitrary units. The instrument was set to acquire in the mass range of most expected cellular lipids and therefore *m/z* 100–1500 was chosen with a mass resolution of 70,000 (defined at *m/z* 200) (*see* **Note 18**).

7. Assess mass accuracy, instrument variability, and extraction reproducibility using exogenous lipid standards that were spiked into samples (*see* **Notes 19** and **20**).

4 Notes

1. At least $3–5 \times 10^6$ cells are needed for the LC/MS analysis of the lipid extracts to obtain an instrument signal response high enough for quantitation. Working with less than 1×10^6 cells in a cell pellet will result in lower lipid sensitivity. As a general sample collection guideline, 3×10^6 Jurkat T lymphocyte cells will yield $1 \times 10^7–1 \times 10^8$ peak intensity on the Thermo Q-Exactive (*with the following parameters: 100 μL reconstitution and 5 μL injection*) for most phospholipids. The smaller the cell size, the higher the cell count required for LC/MS analysis. Pooling multiple batches of lymphocyte cells will increase the overall yield.

2. An ammoniated cell washing buffer (ammonium formate or ammonium acetate) with concentrations as high as 150 mM can be used as a rinsing solvent. Ammonium formate (40 mM or 0.3%) was used for all studies. Ammoniated cell washing buffers maintain the cellular integrity and are compatible with mass spectrometric analysis [14, 18].

3. Scale the stock solution of the re-extraction solvent depending on the number of samples. Because 1 mL of the extraction solvent is added to the aqueous layer, the volume suggested can be applied to 15 samples. Centrifuge sample again after adding the re-extraction solvent to assist in the separation of the aqueous and organic layers.

4. Employing a UHPLC column with an appropriate LC system enables faster analytical run times without compromising overall separation and performance.

5. Water was added to mobile phase B which was originally composed of isopropanol:acetonitrile (90:10, v/v) to aid in the dissolution of 10 mM ammonium formate, resulting in the final composition of isopropanol:acetonitrile:water (90:8:2, v/v).

6. Before centrifugation, normalize samples to the cell concentrations using cell count if available. This step should be performed prior to sample preparation (cell washing or lipid extraction).

7. During the last washing step, reconstitute the cells in the rinsing solution in order to obtain an aliquot for the protein and/or DNA measurements. At this point, cells can also be aliquoted into predetermined amounts prior to lipid extraction.

8. The interpretation of cellular metabolomics data is dependent on the normalization method applied as variations introduced during sample collection, sample preparation, and/or instru-

mental analysis are inevitable. DNA and protein can be used as a means to normalize data post-data acquisition if cell count information is not provided for pre-data acquisition normalization [16]. However, a minimum of $0.5-1 \times 10^6$ cells (5 µL aliquot dissolved in cell washing buffer) should be used to acquire accurate DNA and/or protein measurements for normalization. For these studies, protein concentrations were quantified by fluorescence using a Qubit Protein Assay (Quant-iT Protein Assay Kit, Thermo Fisher Scientific) performed on a Qubit 3.0 Fluorometer. For post-data acquisition normalization to protein levels, the integrated peak area for each feature within a sample was divided by the respective protein concentration.

9. Cells should be stored as a cell pellet at −80 °C until analysis. Perform the lipid extraction on the same day as the data acquisition to avoid oxidation or nonenzymatic degradation of lipid species.

10. Sonicate the lipid internal standard mix (*with heat*) to ensure dissolution of the lipids in the stock solvent before spiking into the sample. Use a repeater pipette to avoid variation and allow for equal aliquots to be spiked into all samples.

11. A minimum of 2 mL of methanol is needed to effectively lyse and extract the intracellular components from one million lymphocyte cells. Butylated hydroxytoluene (1 mM BHT) in methanol can be substituted for 100% methanol to minimize oxidation of various lipid species. Mix the solvent vigorously to break up the cell pellet. Be sure to use a conical tube that is compatible with chloroform (e.g., polypropylene or glass).

12. Vortex every 2–3 min during the lipid extraction to ensure mixing of the solvents and total lysis of the lymphocyte suspension cells.

13. The addition of water aids in the separation of the aqueous (upper) and organic (lower) layers.

14. The protein content from the lymphocyte cells is located between the aqueous and organic layers.

15. The aqueous layer from this extraction step can be isolated via centrifugation, stored in a 15 mL conical tube, and analyzed for metabolomic analysis after reconstitution in water with 0.1% formic acid.

16. Cell lipid extracts reconstituted with isopropanol should be analyzed on the same day or soon after to prevent lipid oxidation and/or degradation during storage.

17. The organic solvents used should be of LC/MS grade. Mobile phases can be stored at room temperature.

18. Under these UHPLC conditions, the isomeric separation of lysoglycerophospholipids can be observed in positive ion mode as shown in Fig. 2. The LPC with the fatty acid tail in the *sn*-2

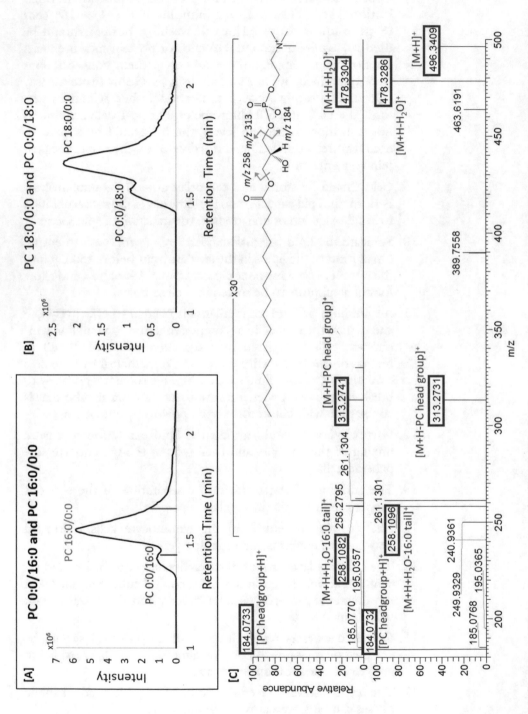

Fig. 2 (a) Co-elution profile of PC 16:0/0:0 (LPC 16:0, sn2) and PC 0:0/16:0 (LPC 16:0, sn1) at *m/z* 496.3398. **(b)** Co-elution profile of PC 18:0/0:0 (LPC 18:0, sn1) and PC 0:0/18:0 (LPC 18:0, sn2) at *m/z* 524.3711. **(c)** Example MS/MS spectra of LPC 16:0 at 30 NCE with the MS/MS spectra of LPC 16:0 (sn2) at 1.44 min and the MS/MS spectra of LPC 16:0 (sn1) at RT 1.64 min

position of the glycerol backbone has a lower ion intensity and elutes first due to a weaker hydrophobic interaction of the branched acyl chain with the C18 stationary phase [19–21].

19. Lipid extracts can be analyzed in negative ion mode to aid in the detection and MS/MS identification of lipid species which preferentially ionize in this mode. The same LC method is used for positive and negative ion modes with only a few modifications of the source condition. Polarity ion switching can be employed to increase the throughput of data analysis by reducing the instrument resolution to 35,000 (defined at m/z 200). Table 1 provides a list of the most abundant lipid species and adducts detected in positive and negative ion modes for T cells. Triacylglyceride (TG) and diacylglyceride (DG) species are detected as an ammoniated adduct in positive ion mode.

Table 1
Most abundant lipid species and adducts detected for mammalian lipid extracts by UHPLC-HRMS in positive and negative ion modes

Lipid class	Abbreviation	Ion detected	
		Positive mode	Negative mode
Cardiolipin	CL	–	$[M-H]^-$, $[M-2H]^{-2}$
Cholesteryl (steryl) ester	CE	$[M + NH_4]^+$	–
Ceramide (N-acylsphingosine)	Cer	**[M + H]+**	–
Glucosylceramide	GlcCer	**[M + H]+**	–
Ceramide 1-phosphate	CerP	$[M + H]^+$	$[M-H]^-$
Diacylglyceride	DG	$[M + NH_4]^+$	–
Lysophosphatidylethanolamine	LPE	**[M + H]+**, $[M + Na]^+$	$[M + HCO_2]^-$
Lysophosphatidylcholine	LPC	**[M + H]+**, $[M + Na]^+$	$[M + HCO_2]^-$
Lysophosphatidylinositol	LPI	–	$[M-H]^-$
Phosphatidylcholine	PC	**[M + H]+**, **[M + Na]+**	**[M + HCO2]−**
Phosphatidylethanolamine	PE	**[M + H]+**, **[M + Na]+**	**[M−H]−**
Phosphatidylglycerol	PG	$[M + H]^+$, $[M + Na]^+$, **[M + NH4]+**	**[M−H]−**
Phosphatidylserine	PS	**[M + H]+**, $[M + Na]^+$	**[M−H]−**
Phosphatidylinositol	PI	**[M + H]+**, $[M + NH_4]^+$	**[M−H]−**
Sphingomyelin	SM	**[M + H]+**	–
Triacylglyceride	TG	**[M + NH4]+**	–

In *bold* are the lipid adducts most commonly found in T cells

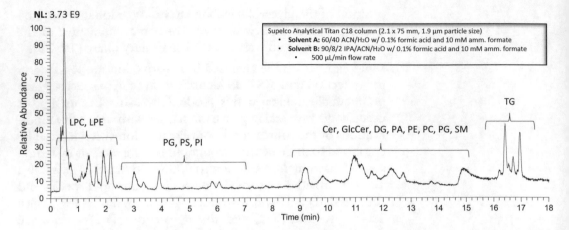

Fig. 3 Total ion chromatogram of the reverse-phase endogenous lipid elution profile for extracted Jurkat T lymphocyte cells in positive ion mode using a Supelco Analytical Titan C18 column

20. The lipid standards chosen elute throughout the chromatogram and the concentration chosen (15 ppm) has a similar magnitude to the peak area of endogenous lipids. Fig. 3 shows a UHPLC-HRMS total ion chromatogram lipid elution profile of a T cell lipid extract acquired in positive ion mode.

Acknowledgment

This work was supported by the Southeast Center for Integrated Metabolomics (SECIM) – NIH Grant #U24 DK097209.

References

1. Bird SS, Marur VR, Sniatynski MJ et al (2010) Lipidomics profiling by high resolution LC-MS and HCD fragmentation: focus on characterization of mitochondrial cardiolipins and monolysocardiolipins. Anal Chem 83(3)940–949. doi:10.1021/ac102598u

2. Wang M, Huang Y, Han X (2014) Accurate mass searching of individual lipid species candidates from high-resolution mass spectra for shotgun lipidomics. Rapid Commun Mass Spectrom 28(20)2201–2210. doi:10.1002/rcm.7015

3. De la Luz-Hdez K (2012) Metabolomics and mammalian cell culture. In: Roessner U (ed) Metabolomics. InTech Rijeka, Croatia, pp 1–17

4. Blanksby SJ, Mitchell TW (2010) Advances in mass spectrometry for lipidomics. Annu Rev Anal Chem (Palo Alto Calif) 3:433–465. doi:10.1146/annurev.anchem.111808.073705

5. Wenk MR (2005) The emerging field of lipidomics. Nat Rev Drug Discov 4(7)594–610

6. Vihervaara T, Suoniemi M, Laaksonen R (2014) Lipidomics in drug discovery. Drug Discov Today 19(2)164–170. doi:10.1016/j.drudis.2013.09.008

7. Dehairs J, Derua R, Rueda-Rincon N et al (2015) Lipidomics in drug development. Drug Discov Today Technol 13:33–38. doi:10.1016/j.ddtec.2015.03.002

8. Marechal E, Riou M, Kerboeuf D et al (2011) Membrane lipidomics for the discovery of new antiparasitic drug targets. Trends Parasitol 27(11)496–504. doi:10.1016/j.pt.2011.07.002

9. Meikle PJ, Wong G, Barlow CK et al (2014) Lipidomics: Potential role in risk prediction and therapeutic monitoring for diabetes and cardiovascular disease. Pharmacol Ther 143(1)12–23. doi:10.1016/j.pharmthera.2014.02.001

10. Zhou X, Mao J, Ai J et al (2012) Identification of plasma lipid biomarkers for prostate cancer by lipidomics and bioinformatics. PLoS One 7(11) e48889. doi:10.1371/journal.pone.0048889

11. Hinterwirth H, Stegemann C, Mayr M (2014) Lipidomics: quest for molecular lipid biomarkers in cardiovascular disease. Circ Cardiovasc Genet 7(6)941–954. doi:10.1161/CIRCGENETICS.114.000550

12. Zhao YY, Cheng XL, Lin RC (2014) Lipidomics applications for discovering biomarkers of diseases in clinical chemistry. Int Rev Cell Mol Biol 313:1–26. doi:10.1016/B978-0-12-800177-6.00001-3

13. Zhao Y, Cheng X, Lin R et al (2015) Lipidomics applications for disease biomarker discovery in mammal models. Biomark Med 9(2)153–168. doi:10.2217/bmm.14.81

14. Ulmer CZ, Yost RA, Chen J et al (2015) Liquid chromatography-mass spectrometry metabolic and lipidomic sample preparation workflow for suspension-cultured mammalian cells using jurkat T lymphocyte cells. J Proteomics Bioinform 8(6)126–132. doi:10.4172/jpb.1000360

15. Dietmair S, Timmins NE, Gray PP et al (2010) Towards quantitative metabolomics of mammalian cells: Development of a metabolite extraction protocol. Anal Biochem 404(2)155–164. doi:10.1016/j.ab.2010.04.031

16. Silva LP, Lorenzi PL, Purwaha P et al (2013) Measurement of DNA concentration as a normalization strategy for metabolomic data from adherent cell lines. Anal Chem 85(20)9536–9542. doi:10.1021/ac401559v

17. Reis A, Rudnitskaya A, Blackburn GJ et al (2013) A comparison of five lipid extraction solvent systems for lipidomic studies of human LDL. J Lipid Res 54(7)1812–1824. doi:10.1194/jlr.M034330

18. Berman ES, Fortson SL, Checchi KD et al (2008) Preparation of single cells for imaging/profiling mass spectrometry. J Am Soc Mass Spectrom 19(8)1230–1236. doi:10.1016/j.jasms.2008.05.006

19. Lee JY, Min HK, Moon MH (2011) Simultaneous profiling of lysophospholipids and phospholipids from human plasma by nanoflow liquid chromatography-tandem mass spectrometry. Anal Bioanal Chem 400(9)2953–2961. doi:10.1007/s00216-011-4958-7

20. Onorato JM, Shipkova P, Minnich A et al (2014) Challenges in accurate quantitation of lysophosphatidic acids in human biofluids. J Lipid Res 55(8)1784–1796. doi:10.1194/jlr.D050070

21. Granafei S, Losito I, Palmisano F et al (2015) Identification of isobaric lysophosphatidylcholines in lipid extracts of gilthead sea bream (*Sparus aurata*) fillets by hydrophilic interaction liquid chromatography coupled to high-resolution Fourier-transform mass spectrometry. Anal Bioanal Chem 407(21)6391–6404. doi:10.1007/s00216-015-8671-9

Chapter 14

Ambient Lipidomic Analysis of Brain Tissue Using Desorption Electrospray Ionization (DESI) Mass Spectrometry

Valentina Pirro, Alan K. Jarmusch, Christina R. Ferreira, and R. Graham Cooks

Abstract

Desorption electrospray ionization (DESI) is a spray-based ambient ionization method for mass spectrometry (MS) which generates ions in native atmospheric conditions (e.g., pressure and temperature). Ambient ionization allows in situ analysis of unmodified biological samples by eliminating analyte extraction and separation steps before MS analysis. DESI-MS has been extensively used to analyze organ tissues both in humans and in vertebrate animals, focusing on the detection of small molecules (e.g., oncometabolites, xenobiotic drugs, hormones, etc.) and lipids.

DESI-MS lipidomic analysis workflow involves (1) detection of lipids from intact biological material, (2) detection and identification of lipids in complex mixtures, and (3) discrimination between similar lipids, e.g., isomeric lipids. DESI-MS can provide lipid profiles using a relatively fast and simple workflow in which low-resolution single-stage mass spectra are recorded during 2D or 3D image analysis (i.e., mapping the distribution of lipids within the sample). Such DESI-MS lipid profiles include many classes of lipids, such as phosphatidylcholines (PC), triacylglycerols (TGs), free fatty acids (FFAs), phosphatidylethanolamines (PEs), phosphatidylinositols (PIs), phosphatidylserines (PSs), diacylglycerols (DGs), ubiquinone, and cholesterol derivatives (e.g., cholesterol sulfate and cholesterol esters). Depending on the mass spectrometer used, there is the additional possibility of obtaining structural information of lipids via MS^n, and molecular formulae via high resolution mass spectrometry (HRMS). Focusing on the analysis of human brain, here we summarize the DESI-MS experimental workflow for tissue analysis, data collection, and processing using low and high-resolution mass spectrometers, emphasizing the strategies for structural identification of lipids.

Key words DESI-MS, Reactive DESI, Ambient mass spectrometry, Multidimensional MS scan, Exact mass measurement, Lipid, Biopsy, Tissue section, Human brain, Mouse brain

Electronic supplementary material: The online version of this chapter (doi:10.1007/978-1-4939-6946-3_14) contains supplementary material, which is available to authorized users.

Paul Wood (ed.), *Lipidomics*, Neuromethods, vol. 125,
DOI 10.1007/978-1-4939-6946-3_14, © Springer Science+Business Media LLC 2017

1 Introduction

Ambient ionization coupled to mass spectrometry (MS) refers to the generation of ions from unmodified samples outside the vacuum system of the mass spectrometer (e.g., under native temperature and pressure), followed by ion transfer and analysis in the MS system [1–3]. No sample handling and analysis speed are primary advantages of ambient ionization [4]. Desorption electrospray ionization (DESI) is the first ambient method that was reported over a decade ago. Since then, the ambient methods have expanded into a large portfolio with more than a dozen reported methods [3, 5]. DESI is an experiment, in which a spray of charged solvent is directed onto a sample (e.g., thin tissue section or smear, a dried biofluid spot) with a spot size that typically ranges between 50 and 200 μm. A thin liquid film is formed on the sample surface, dissolving soluble compounds, and when subsequent droplets impact the surface, they sputter smaller secondary droplets containing compounds derived from the surface. The secondary droplets are sampled by the mass spectrometer, and gas-phase ions are generated via electrospray-like mechanisms. 2D DESI-MS imaging experiments are performed by rastering the sample under the spray spot in a controlled fashion [6].

DESI has been applied for the detection of diverse molecules, including small metabolites [7], drugs [8], chemical agents [9], and lipids [10] in human and animal tissue organs, cell pellets, biofluids, oocytes, and embryos. Physicochemical properties of the target molecules affect the conditions to be used for the DESI-MS analysis, both in terms of choice of DESI solvent and modality of ion acquisition for the MS system (e.g., positive or negative ionization mode, single or multistage MS). The capabilities of DESI for tissue and microscopic samples (such as oocytes and preimplantation embryos) analysis have been significantly improved by the use of nondestructive (or histologically compatible) solvent systems that allow chemical information to be obtained, while preserving sample morphology for subsequent histochemical or other identifications [11]. Specifically for embryo analysis, histologically compatible solvents such as dimethylformamide—acetonitrile (DMF—ACN, 1:1 v/v) can be used to gently extract the lipids from microscopic oocytes and embryos, while providing signal for many seconds so that mass spectra can be accumulated and averaged to generate a representative lipid profile [12–14]. For tissue analysis, samples analyzed by histologically compatible DESI-MS can be stained with hematoxylin and eosin (H&E) after analysis, enabling direct correlation between molecular and morphological information [15].

Lipidomic studies are of interest for tissue and embryonic analysis because lipids play essential roles in cells as signaling molecules, membrane components, and energy stores, and changes in lipid metabolism are associated with cell differentiation and with a number of diseases including cancer, diabetes, and obesity [16]. Exploring how lipids are regulated is the key to understanding biological pathways, organogenesis, and developmental processes occurring in a biological system. To date, DESI has been used to study prostate [17], bladder [18], kidney [19], gastrointestinal [20], lymphoma [21], pancreatic, breast, and brain cancers. Specifically for brain, the lipid patterns recorded by DESI allowed differentiation of normal dura matter from meningioma, pituitary, and glioma cancers [22]. Lipid differences have been explored among glioma subtypes, grades, and tumor cell concentrations (i.e., relative percentage of tumor compared with parenchyma) [23, 24]. Normal gray matter is comprised of glia and unmyelinated neurons; DESI negative mass spectra are dominated by phosphatidylserine PS(40:6), with predominant acyl chains 18:0 (stearic acid) and 22:6 (docosahexaenoic acid), as attributed by high-resolution (HR) MS and low-resolution MS^n data. Positive ion mode data show greater abundance of potassiated adducts of phosphatidylcholines (PCs). Normal white matter is characterized by PS(18:1_18:0) and sulfatides, the detection of which correlates with the increased presence of myelinated neurons, and increased abundance of potassiated and sodiated adducts of galactosylceramides, which are the principal glycosphingolipids in brain tissue. Mass spectra showing mixtures of all these lipids are common in normal brain samples, reflecting the typical parenchyma composition that is a mixture of glial cells, unmyelinated and myelinated neurons [22]. Cancerous tissue has a drastic reduction of those lipids in the DESI positive and negative spectra but increased abundance of phosphatidylinositol PI(18:0_20:4), as well as the chlorinated and sodiated adducts of PCs. Such an increase is consistent with ^{1}H NMR data [25] and can be related to the role of PCs in cell proliferation which is increased in cancer [26]. Interestingly, changes in the relative abundance of chlorinated, potassiated, and sodiated adducts for the same lipids occur reproducibly between normal and cancerous tissues. The measured differences could be due to either altered concentration of the adducting species (e.g., cations) in the tissue, different concentrations of the lipids themselves, or matrix effects that cause differences in the ionization efficiency. Note, for example, that lipids can aggregate depending on their concentration, hydrophobicity, etc., and this phenomenon can affect ionization. Squalene, ubiquinone (coenzyme Q10), several triacylglycerols (TGs), and diacylglycerols (DGs) have been detected by DESI-MS in human and mouse brain tissue (unpublished data). Cytosolic lipids play an essential role for energy storage and mitochondrial activity (e.g., coenzyme Q10 is a fundamental electron carrier in the mitochondrial respiratory chain) and are known to be altered with metabolic diseases, including cancer [27]. Note that lipid changes detected by DESI-MS are downstream metabolic consequences of upstream events (gene expression and protein activity) and are also affected by degradation, turnover, and de novo synthesis. The exploratory DESI-MS imaging studies on frozen mouse and human brain tissue sections contributed to the general knowledge of the biochemistry of normal brain and its alteration in cancer [22–24]. Such studies have paved the way for the development of a rapid intraoperative DESI-MS analysis of fresh brain tissue smears, which might serve as a molecular-based cancer diagnostic tool [22].

Here, we describe and comment on the procedure of performing lipidomic analysis of brain tissue by DESI-MS. DESI conditions and materials are listed. The classes of lipids detected using different solvent systems are described as well as procedures to identify the lipid structures using low-resolution MS^n via low-energy collision-induced dissociation (CID), and HRMS. The present work is limited to the analysis of lipids in brain tissue by DESI-MS and is not intended to be comprehensive from a lipidomic point of view; however, the DESI-MS methodology presented here is applicable to the analysis of other organs and types of biological samples.

2 Materials

Reagents should be at least analytical grade (ideally MS grade) and prepared daily at room temperature unless otherwise stated. Chemicals should be used as received without further purification. The solutions for the DESI spray are usually prepared in glass vials and sonicated for a few minutes before usage, paying attention not to have contact between the solvent and the vial cap. Vials and tubes that leach plasticizers, polymers, and coating materials based on the solved used should be avoided. Please follow chemical and biohazard waste disposal regulations after sample analysis.

2.1 Samples

1. Snap-frozen human brain tissue and mouse brain specimens (*see* **Note 1**).

2. Fresh human brain biopsy tissue (*see* **Note 2**).

2.2 Materials and Reagents

1. Superfrost® Plus microscope glass slides (*see* **Note 3**).

2. Black marker (*see* **Note 4**).

3. Glass vials, e.g., Kimble® 60910L-1 Clear Glass Sample Vial with Phenolic Cap & Rubber Liner, O.D. x Height: 15 × 45 mm.

4. Acetonitrile (ACN).

5. Dimethylformamide (DMF).

6. Silver nitrate, $AgNO_3$ (*see* **Note 5**).

7. Compressed nitrogen.

8. Smearing device (*see* **Note 6**).

9. H&E staining reagents. The description of the H&E procedure is beyond the scope of this chapter. For further information, *see* refs. 6, 22.

2.3 Mass Spectrometry Instrumentation

1. DESI-spray, DESI source, and moving stage (*see* **Note 7**).

2. Mass spectrometers (data presented here were acquired using the Exactive Orbitrap mass analyzer and LTQ low-resolution linear ion trap of Thermo Scientific (San Jose, CA)). The only modifications to the instruments were the installation of a custom DESI source, which included a source override adapter, an external cable for the application of the high voltage, and an extended stainless-steel ion transfer capillary (total length, 180 mm, inner diameter, 0.02″, outer diameter, 1/16″, length protruding from the MS vacuum system, ~87.5 mm). Swagelok stainless-steel fittings were used to connect the capillary with the threaded MS inlet. The 1/16″ Swagelok fittings were machined so that the capillary can be placed all the way through the fitting and secured on one side with a ferrule and metal nut.

2.4 Other Instrumentation

1. Ultra-Low Temperature Freezer (−80 °C).

2. Cryotome; optical cutting temperature (OTC) compound; low-profile disposable microtome blades and paintbrushes (*see* **Note 8** and ref. 6 for additional details).

3. Desiccator (*see* **Note 9**).

4. Syringe pump and compatible glass syringes (*see* **Note 10**).

5. Caliper.

3 Methods

3.1 Frozen Sample Handling

The detailed procedure on how to prepare tissue sections for DESI-MS analysis is reported in [6]. Briefly:

1. Place the frozen brain tissue on the sample holder of the cryotome using minimal amount of OCT.

2. Cut the tissue into 5–15 μm sections. Optimize the cryotome temperatures following standard protocols for brain tissue.

3. Collect tissue section using a paintbrush and place it on the microscope glass slide avoiding folding or cracking the tissue, and paintbrush effects.

4. If DESI-MS analysis is not immediately performed, place the glass slides with samples in a slide box and store it at −80 °C until analysis.

3.2 Fresh Sample Handling

1. Place the fresh brain biopsy tissue (about 10–50 mm³) on the microscope glass slide, preferably in the center of the slide, close to the labeling area.

2. Position the smear device on the glass slide and smear the tissue across the entire surface of the glass, by gently pressing down while holding the slide on a hard surface (*see* **Notes 11** and **12**). Smear multiple times if necessary and avoid leaving aggregates of tissue on the slide; irregular height of the excess material could damage the DESI spray or contaminate the extended inlet capillary during analysis. An example of a tissue smear is shown in Fig. 1.

3. If DESI-MS analysis is not immediately performed, place the glass slides with samples in a slide box and store it at −80 °C until analysis.

3.3 DESI Spray Solvent Preparation

1. Spray solution for conventional DESI-MS analysis in negative and positive ion mode ionization: DMF—ACN (1:1, v/v) (*see* **Note 13**).

2. Spray solution for reactive DESI-MS analysis in positive ion mode ionization: ACN with 3.0–5.0 ppm silver nitrate (*see* **Notes 14** and **15**).

3.4 DESI-MS Optimization

1. Optimize the DESI spray to obtain a small slightly elliptical and stable spray spot. *See* refs. 6, 13 for additional details on how to build and optimize a DESI spray.

2. Direct the DESI spray onto the top of a glass slide and optimize the spray orientation (e.g., incident angle, distance between the spray tip and the sample surface, distance between the spray tip and the MS inlet) to obtain maximum and stable signal intensity, and a small sampling spot [6]. Typical conditions used are: incident angle, 52°; tip-to-source distance, about 2 mm; tip-to-inlet distance, about 4 mm (*see* **Note 16**).

3. Optimize the flow rate of the DESI spray solvent system and pressure of the nitrogen gas. For conventional DESI analysis

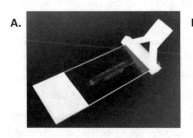
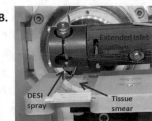

Fig. 1 (**a**) Image of a 3D printed device used to smear fresh brain tissue across a microscope glass slide. (**b**) DESI-MS interface with custom slide holder for rapid intraoperative analysis of freshly resected human brain tissue smears. (**c**) Brain tissue smeared with a 3D printed device and H&E stained after DESI-MS analysis

using DMF—ACN (1:1, v/v), the flow rate is usually between 0.7 and 1 µL/min. For reactive DESI analysis using ACN doped with $AgNO_3$, the flow rate is usually between 4 and 6 µL/min. Nitrogen gas is commonly set to values between 150 and 170 PSI. These parameters should be altered on the manufacturer's tolerances as well as the solvent used and other conditions.

4. Optimize capillary temperature and high voltage applied to the syringe needle. Common values are 275 °C and 4.5–5 kV, respectively, for both positive and negative ion mode ionization.

5. Tune the mass spectrometer using an analytical standard, ideally the molecule of interest or one having a similar mass-to-charge (m/z) ratio. The tuning procedure optimizes ion optics and transmission quadrupoles or hexapoles (depending on the instrument) potentials to maximize ion transmission (*see* **Note 17**). Tuning can also be done with an alternative ionization technique, such as nanoelectrospray ionization.

6. Optimize injection time and the number of microscans to produce a stable and high-quality MS signal intensity throughout the mass range of interest, without space charging (deviation in ion trajectories which impact ion m/z values and resolution in the mass spectra. This phenomenon is most noticeable in ion trap instruments that confine large numbers of ions in a small volume). This is important, especially if DESI-MS images are run with the automatic gain control (AGC) set off.

3.5 DESI-MS Analysis

1. For the analysis of frozen samples, remove the glass slide containing the samples from the freezer and place it in the vacuum desiccator for a few minutes, until all visible water has been removed. Conversely, fresh tissue smears can be analyzed immediately by DESI-MS.

2. When running a DESI-MS image, measure the sample with a caliper to calculate the area to be imaged. Position the glass slide under the DESI spray on the moving stage optimized as described above. Start MS analysis and data acquisition. All these steps have been thoroughly described in reference [6]. When rapidly analyzing tissue smears, researchers can acquire either a DESI-MS image of the whole smear or a portion of it, or acquire just a couple of scan lines across the tissue, to recover an average mass spectrum representative of the smear. Preliminary DESI imaging data proved that the act of smearing is sufficient to homogenize the tissue [21, 22]. An overview of the whole experimental workflow of DESI-MS imaging of tissue sections versus a rapid DESI-MS analysis of tissue smears is illustrated in Fig. S1.

3. When using nondestructive solvent systems (*see* **Note 18**), the same tissue section or smear can be analyzed multiple times using different DESI-MS conditions (e.g., different solvent systems or ionization mode) [22, 28, 29], so that a broader range of lipid classes can be detected (*see* **Note 19**). The moving stage can be automatically reset to the origin position after the first image or when the rapid DESI-MS analysis is completed, in order to have superimposable spatial information, provided that the duty cycle of both analyses is kept the same.

4. Once the MS data acquisition is complete, mass spectra can be visualized using the instrument software, like XCalibur for Thermo Scientific instruments. Representative DESI-MS spectra for normal human gray and white matter, glioma tissue, and the ion images of a human brain biopsy specimen for lipids PS(40:6) and PS(36:1) (respectively, deprotonated ions of m/z 834.5 and 788.5) are illustrated in Fig. 2. Raw files can be exported as different file formats (e.g., excel and text) or converted with open source software (e.g., into mzXML files with protewizard MS converter—http://proteowizard.source-forge.net—or hdr files with Analyze 7.5 which is required by Biomap). DESI images can be visualized with the software Biomap (http://www.maldi-msi.org). Ref. 6 describes in detail each step necessary to import and analyze the data. Other software also implements specific imaging tools for the

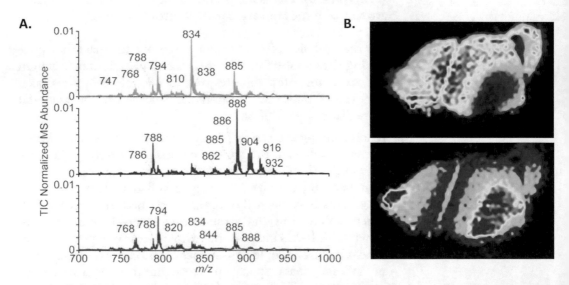

Fig. 2 (a) DESI-MS spectra in negative ion mode for human normal gray matter (*green, N* = 223), normal white matter (*blue, N* = 66), and gliomas (*red, N* = 158). Copyright from Jarmusch et al. *Proc Natl Acad Sci U S A.* (2015) 113:1486–1489. **(b)** DESI-MS ion images for ions of m/z 834.5 (*top*) and 888.5 (*bottom*) from a human brain frozen tissue section. The different distribution of the ions within the tissue is associated with the fact that the tissue is predominantly from gray matter (i.e., higher intensity of 834.5), with two small pockets of white matter (i.e., higher intensity of 888.5), as confirmed by histopathology

analysis of MS data (e.g., MSiReader, Matlab, Mathworks, http://www4.ncsu.edu/~dcmuddim/msireader.html) [30]. For tissue smears, the mass spectra can be averaged by selecting across multiple scans acquired over time, and the data exported as a list of m/z values and ion abundances. Information on how to visualize and process DESI-MS (imaging) data is beyond the scope of this chapter. Several articles report extensive information on how lipid profiles collected by DESI-MS are analyzed with pattern recognition methods [12, 20, 22, 23, 28, 29, 31]. For the sake of simplicity, here we will focus on lipid attributions based on low-resolution and low-energy CID MS^n and HRMS measurements.

3.6   Lipid Attribution

3.6.1   Nomenclature and Structural Resolution

The use of a correct and updated nomenclature system when annotating lipids detected by MS is fundamental to allow for universality in literature and database searches, and to ease exchange of information, especially considering the number of lipid classes and subclasses, and the fact that lipids exist in a variety of isobaric, structural isomeric, and stereoisomeric forms. Several publications [32, 33] and online sources (e.g., http://www.lipidmaps.org/data/classification/LM_classification_exp.php) describe a comprehensive classification system for lipids. Efforts have also been made in creating and standardizing a practical shorthand notation of lipids structures derived from MS approaches that enables correct and concise reporting of data [34]. Briefly, a shorthand annotation for lipids will show (1) an acronym to identify the lipid class (e.g., PC for phosphatidylcholine), (2) the number of carbons and unsaturations (C:U) of the lipid components, such as fatty acids, either as total sum (e.g., PI(36:1)) or listing the individual constituents (e.g., PC(18:1_18:0)), (3) specific notations and symbols when isomeric (e.g., sn-positional isomers, double-bond isomers) and stereoisomeric information (e.g., *cis, trans*) is known (e.g., PC(16:0/18:1, *n-9*, *cis*)). Essentially, the MS analysis determines the level at which a lipid can be determined. This differs greatly with the type of MS instrumentation and methodology adapted (e.g., an MS/MS low-resolution mass spectrometer coupled with chromatography versus shotgun HRMS; use of derivatization systems prior to MS analysis, etc.). Table S1 shows an example of how the nomenclature can be refined as the MS analysis becomes more and more structurally specific.

3.6.2   Low-Resolution and Low-Energy CID MS^n Fragmentation
(See Note 20*)*

1. Collect the product ion-scans DESI-MS^n data [12] (*see* **Note 21**) from one or more precursor ions of interest by isolating each ion sequentially using the scan mode settings in the mass spectrometer software (*see* **Note 22**). Data can be acquired by spotting a small area of a tissue section or smear and averaging scans over time, or by recording an entire DESI-MS/MS image to visualize spatial distributions of product ions (*see* **Note 23**). A precursor ion can be of interest because (1) it

relates to an expected compound, (2) it has been targeted a priori, or (3) it appears discriminatory between classes of samples, e.g., ions present only in the mass spectra of normal white matter or normal gray matter samples (Fig. 2).

2. After isolation of the target precursor ion, select CID as the type of fragmentation and increase the collision energy in a stepwise fashion—from low to moderate to high—to observe product ions. The amount of collision energy required to observe ion fragmentation is related to the stability of the ion. Product ions of greater intensity are indicative of more thermodynamically stable structures which result from breaking weaker bonds in the precursor ion. It is common practice to choose conditions that retain 5–10% relative abundance of the precursor ion, while yielding product ions with stable intensities and consistent ratios between them. The Mathieu q parameter can be optimized as well in the same manner.

3. If abundant products are detected in the MS^2 experiment, an additional isolation and fragmentation step can be performed (just proceeding as described above) to collect MS^3 sequential product ions ($\bullet \rightarrow \bullet \rightarrow \circ$) [12] and gain further structural information.

4. When performing ambient MS analysis on unaltered tissue sections and smears, it is common to detect lipids as chlorinated $[M+Cl]^-$ (negative ion mode), sodiated $[M+Na]^+$ and potassiated $[M+K]^+$ (positive ion mode) adducts, together with deprotonated $[M-H]^-$ and protonated $[M+H]^+$ molecular species, due to the endogenous presence of chlorine, sodium, and potassium in brain. In other cases, the formation of adducts is achieved by doping the DESI spray with specific metals or salts, e.g., reactive DESI experiments with silver nitrate. The presence of adducting species can provide additional information for lipid identification by MS^n [35–39] (see Note 24). For example, isotopic peaks can be seen in the full scan mass spectra (e.g., $[M+^{35}Cl]^-$ and $[M+^{37}Cl]^-$ with a relative intensity ratio of approximately 3:1; $[M+^{107}Ag]^+$ and $[M+^{109}Ag]^+$ with a 1:1 relative intensity ratio) and they can be isolated and fragmented individually to provide confirmatory structural information. Indeed, the fragment ions retaining the adducting species should be similar but shifted in mass-to-charge (m/z) values when fragmenting the lighter compared to the heavier isotope adducts.

5. Collect the product ion scan of the precursor ions of interest and average several scans over time. Compute the neutral losses by subtracting the m/z value of the observed product ions

from that of the precursor ion or another product. Be aware that CID spectra can show product ions coming from multiple isobaric species fragmented simultaneously, as no separation of molecules occurs before DESI-MS analysis, which makes lipid identification more challenging.

6. To further confirm lipid annotations based on low-energy CID DESI-MS^n data, pure analytical standards can be purchased and analyzed under the same conditions, or literature as well as database spectra can be compared with the experimental results.

3.6.3 Full-Scan High-Resolution Mass Measurements

1. In our laboratory, it is common practice to collect DESI-HRMS data to complement the MS^n results and increase our confidence in lipid identification. DESI-HRMS experiments use the same conditions as described above and settings of the MS system are optimized in order to get high-quality signal (e.g., injection time, tuning, resolving power, etc.). Calibration of the HR mass spectrometer should be performed before any data are collected. DESI-HRMS data are usually acquired positioning a tissue section or smear under the DESI spray and averaging signal over time; however, DESI-HRMS imaging experiments can be performed as well [40].

2. Collect the spectra and visualize the average scan with the XCalibur software. Record the *m/z* values with four to five decimal digits.

3. Input the measured mass into the search function of the LIPID MAPS or METLIN databases (*see* **Note 25**). Select the tolerance for the mass error in ppm (*see* **Notes 26** and **27**) and the charge state (e.g., positive or negative). Several molecular species and adducts can be selected to limit the search, but this information is not necessarily known or foreseeable. The pattern of peaks detected in HRMS could be indicative of multiply charged ions or isotopic ions; such evidence can guide the search as well. Also, if the DESI spray has been doped with specific metals (e.g., Li) or salts (e.g., ammonium acetate), corresponding adducting species can be selected to ease the search. This search will provide a list of possible molecular formulae (*see* **Note 28**).

4. Copy a molecular formula of interest into the XCalibur function and plot the theoretical isotopic distribution for further confirmation (*see* Fig. 3 and **Note 29**). This step is important especially when complex isotopic patterns are detected, as for the case of silver nitrate adducts. Chemical formulae with non–matching isotopic patterns can be excluded (Fig. S2).

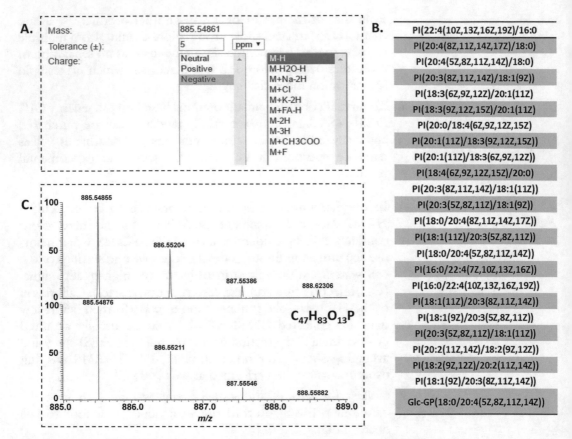

Fig. 3 (a) Screenshot of METLIN simple search function. (b) List of possible lipid identifications based on the search function for the experimental mass measurement of m/z 885.54861 (deprotonated ion, error in mass < 5 ppm), all matching the chemical formula $C_{47}H_{83}O_{13}P$. (c) Comparison between the experimental isotopic distribution for the ion m/z 885.54861 and the theoretical isotopic distribution for deprotonated lipid ions with chemical formula $C_{47}H_{82}O_{13}P$

5. To further confirm lipid annotations based on HRMS data, pure analytical standards can be purchased and analyzed under the same DESI-HRMS conditions, or literature data as well as database or theoretical spectra can be compared with the experimental results.

6. The resolving power of the HR mass spectrometer might not be high enough to completely deconvolute peaks in the mass spectra. Convolution of peaks has been frequently seen in reactive DESI-HRMS experiments with silver nitrate because of the peculiar isotopic pattern of the molecular species $[M+Ag]^+$ and $[M+Ag_2NO_3]^+$. Also, several TGs, DGs, CEs are endogenously present in the brain tissue and they can differ by only 2 m/z units from one another for the presence of one unsaturation more or less. As an example, Fig. 4 shows DESI-HRMS spectra of ubiquinone and ubiquinol as pure standards and in a mixture as detected from a pellet of breast cancer cell line.

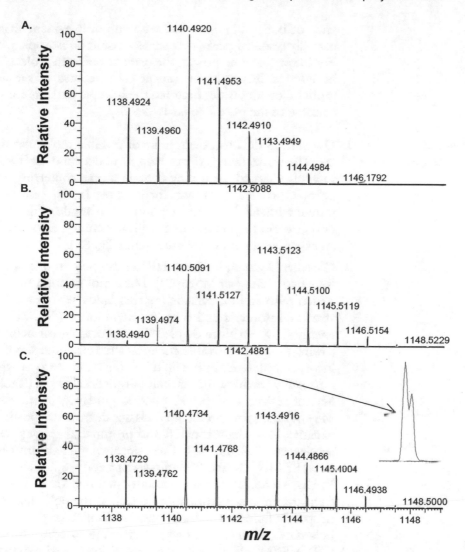

Fig. 4 (**a**) Isotopic distribution of pure standard of ubiquinone detected by DESI-HRMS as silver nitrate adduct $[M+Ag_2NO_3]^+$ using an Orbitrap mass spectrometer with FT resolution of 30,000. (**b**) Isotopic distribution of pure standard of ubiquinol detected by DESI-HRMS as silver adduct $[M+Ag_2NO_3]^+$ using an Orbitrap mass spectrometer with FT resolution of 30,000. (Accept**c**) Isotopic distribution of ubiquinone and ubiquinol detected in mixture from a breast cancer cell line by DESI-HRMS using an Obitrap mass spectrometer with FT resolution of 30,000. The inset shows the zoomed-in area for the peak around m/z 1142.5 with FT resolution of 100,000

Data have been acquired with ACN doped with silver nitrate. Molecular species are $[M+Ag_2NO_3]^+$. Ubiquinol is the reduced form of ubiquinone. The mitochondria energy production system is based on the exchange of electrons between ubiquinone and ubiquinol and the balance between these two molecular species is required for the correct cell functioning. DESI-HRMS spectra have been acquired with different resolving powers using the same Orbitrap mass spectrometer (Thermo Scientific). Only with the highest resolving power, the presence of both ubiquinone and ubiquinol is evident from the partially resolved peaks that share a common isotopic pattern. For lower resolving power, the presence of both molecules can be inferred in the breast cancer cells because of the altered ratios of ion intensities between the isotopic peaks as compared to those of the pure standards.

3.6.4 Structural Identification of Lipids

1. Use HRMS data to obtain a list of possible molecular formulae. The peaks detected can help in understanding if ions are multiply charged, or if they contain atoms conferring specific isotopic profiles, as shown for example in the case of silver nitrate adducts detected in positive ion mode (Fig. 4). This evidence can help filter out chemical formulae associated with an experimental mass measurement (Fig. S2).

2. Considered alone, full-scan HRMS can help in distinguishing isobaric species (*see* **Note 30**), but is not enough to assign a structure to an ion, considering that lipids exist in a variety of structural isomeric and stereoisomeric forms (Fig. 3b). Low-energy CID (*see* **Note 31**) MS^n data provide great help in narrowing down the possible chemical structures and significantly increase confidence in the lipid identification, even though not necessarily reaching the "structurally defined" level (Table S1). Several articles and books provide guidance in interpreting MS/MS data for specific lipid classes detected as positive and negative ions, deprotonated and protonated species, reactive species, etc. [35, 41, 42]. There are also online resources such as METLIN (https://metlin.scripps.edu) fragment search. Briefly, searching for fragments or neutral losses that result from common structural motifs, such as the polar headgroups of specific lipid classes (e.g., phosphocholine or phosphoethanolamine), is an efficient way to start interpreting low-energy CID MS/MS data. Common neutral losses and product ions indicative of specific lipid classes are listed in [35, 41] (*see* **Note 32**). In negative ion mode, the fatty acid residues of most complex lipids (e.g., glycerophospholipids, TGs) can be also detected in the lower mass range (about m/z 200–350). This information helps eliminate some of the possible combinations of fatty acids giving a specific sum of carbon atoms and unsaturations. See, for example, the product ion scan for the deprotonated ions m/z 788.5 and 885.5 (Fig. 5). The HR measurements, the neutral loss $[M-H-87]^-$ (loss of phosphatidylserine [35]), and the product ion of m/z 241 (i.e., cyclic anion of inositol phosphate [35]), respectively, for the two ions, would suggest PS(36:1) and PI(38:4). For the ion of m/z 788.5, the unique fragments at m/z 281.3, 283.3 are indicative of PS(18:0_18:1) among all possible combinations of fatty

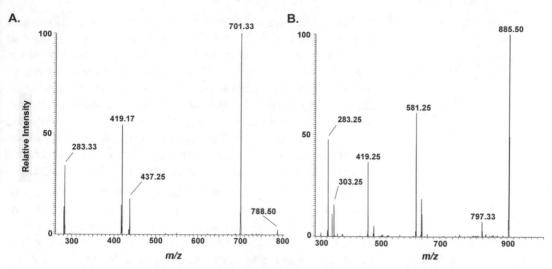

Fig. 5 (**a**) Product ion scan of the ion *m/z* 788.5 acquired from a human brain tissue section by DESI-MS/MS using a low-resolution linear ion trap and low-energy CID. The neutral loss [M−H−87]⁻ is indicative of the PS class. Peaks at *m/z* 281 and 283 correspond to [RCOO]⁻ ions of oleic and stearic acids, suggesting the identification PS(18:0_18:1). Peaks at *m/z* 417 and 419 correspond to the losses of stearic and oleic acid from [M−H−87]⁻ respectively. (**b**) Product ion scan of the ion of *m/z* 885.5 acquired from a human brain tissue section by DESI-MS/MS using a low-resolution linear ion trap and low-energy CID. Peaks at *m/z* 283.3 and 303.3 correspond to [RCOO]⁻ ions of stearic and arachidonic acids, suggesting the identification PI(18:0_20:4). Ion of *m/z* 581.3 corresponds to the loss of arachidonic acid from [M−H]⁻

acids (at least this is the only combination detected by DESI-MS/MS). For the ion of *m/z* 885.5, the fragments at *m/z* 283.3 and 303.3 are indicative of PI(18:0_20:4). This result is not surprising as most mammalian cell membranes include as major constituents PI species containing arachidonate [35]. Note that even carbon chains FAs should be considered predominant and more likely to occur in eukaryotic samples, while odd carbon chain lengths are uncommon. Also, the fatty acyl residues of complex lipids detected by DESI-MS/MS are complementary in information to that of free FAs detected in full-scan DESI-MS, as the soft DESI ionization method does not normally give in-source lipid fragmentation.

3. MS/MS fragments can be listed into the LIPID MAPS software (http://www.lipidmaps.org/tools/), specifying ion intensities for precursor and product ions, and the charge state for the precursor ion, to attempt lipid identification.

4. Positive and negative ions detected for the same lipids (e.g., PCs as chlorinated and potassiated adducts) can be complementary to each other and increase confidence in the identification.

5. Additional structural information to distinguish lipid isomers may be found in MS^n data, but caution is necessary not to over-interpret the experimental evidence, especially when using direct MS methods for mixture analysis. For example, the relative intensity between MS^2 and MS^3 fragments (using the same experimental conditions) might be different between sn-positional isomers for energetic differences [35–39, 41, 42]. Reactive experiments (e.g., Li cathionization) can help in distinguishing double-bond isomers [35]. Stereoisomeric information is not usually recovered from MS^n data instead.

6. Of course, coupling MS^n with orthogonal techniques can provide additional information for lipid identification. For example, chromatography coupled with MS is a leading technique to deconvolute mixtures of molecules and increase the specificity of the MS analysis; however, these solutions are beyond our interest and are not feasible for direct and rapid analysis of complex samples, as is sought by DESI. Ion mobility is an attractive alternative as isomers and stereoisomers can have different cross-sections and ion mobility, and they can be separated in time before MS detection [43, 44]. The coupling of DESI and ion mobility MS has been investigated to detect multiply sialylated ganglioside species, directly from the murine brain tissue. Poly-sialylated gangliosides were imaged as multiply charged ions using DESI, while they were clearly separated from the rest of the lipid classes based on their charge state using ion mobility [45].

4   Notes

1. The tissue should be frozen as promptly as possible after collection to avoid deleterious effects, e.g., physical damage, morphological distortions, and chemical alteration due to enzymatic activity, oxidation, etc. Snap freezing with liquid nitrogen is necessary because slow freezing will cause damage to tissue morphology due to ice crystal formation. Indeed, by snap freezing the water remains in a vitreous form rather than forming crystals that expand when solidified.

2. Microsurgical resection of brain tissue is the primary treatment option for brain tumors. Biopsy is typically used to remove brain tissue. The amount of sample used to prepare tissue smears for DESI-MS analysis is <50 mm³.

3. The use of clean glass is generally preferred to reduce background signal and limit contamination due to extraction of polymers or other coating materials; however, Superfrost® Plus slides are commonly used because they have increased

retention of tissue sections and background observed is minimal.

4. Large-point Sharpie® markers are recommended. Color the back side of the glass slide after the tissue section has been thaw-mounted so that the DESI spray spot can be seen while optimizing and running DESI-MS experiments. The colored side of the glass slide faces the stage sample holder and not the spray [6].

5. Silver nitrate should be of high purity. Reduction to silver metal and formation of silver nanoparticles will occur over time with exposure to air and light. This should be avoided as it results in MS signal instability, increased background, and reduced efficiency in cationization. It is recommended to prepare a 3.0–5.0 ppm solution of $AgNO_3$ in ACN by serial dilution, daily, and to sonicate each solution for a couple of minutes before transferring aliquots of solvent.

6. Smears are commonly prepared for cytological diagnosis. Fresh tissue is spread thinly and unevenly across a microscope slide, usually with another glass slide, to obtain a cellular monolayer that can be stained for histopathology or cytology. A disposable customized 3D printed device is proposed as a more efficient smearing device for DESI-MS analysis, because it is designed to smear the tissue along a glass slide with a fixed width, so that the same rastering program for the DESI moving stage (i.e., area to be analyzed) can be used for all samples. The biopsy can be smeared with a greater thickness and more homogenously, both of which are beneficial for MS signal intensity and stability. Details on the design of the 3D printed device are reported in [22].

7. Commercial DESI sources are available from Prosolia Inc., Indianapolis (http://www.prosolia.com/).

8. OCT should be used to attach the sample to the holder of the cryotome. Complete embedding of the tissue into OCT should be avoided, because the polymer components of the OCT are readily ionized in the positive ion mode.

9. Frozen tissue sections or smears need to be dried before DESI-MS analysis if stored at −80 °C to remove frozen condensation. If a desiccator is not available, samples can be simply left at room temperature until dry.

10. Most commercial mass spectrometers are equipped with a syringe pump. Compatible glass syringes are recommended to deliver the solvent system. Syringes of 250 or 500 µL are appropriate for the flow rates usually adopted in a DESI-MS experiment. Ideally, the syringe should be filled with enough solvent to run an entire set of DESI images.

11. Avoid applying excessive force while smearing the tissue; too much force may completely destroy the cell and render cytology after DESI-MS analysis impractical.

12. Preliminary results showed that the act of smearing sufficiently homogenizes the tissue, which allows one to acquire ions using DESI just by rastering a few times across the tissue smear and then averaging the MS scans, so obtaining a representative mass spectrum for each sample. Correlation analysis performed on DESI-MS data collected on tissue sections and smears prepared from the same specimens confirmed that the chemical information recorded in both cases is similar. Hence, the diagnostic ions detected by DESI-MS imaging on frozen tissue sections are maintained when analyzing fresh tissue smears [21, 22].

13. The main lipid classes detected by DESI-MS in brain tissue as negative ions are fatty acids (FAs), dimers of fatty acids that are formed as gas-phase ions during the desorption/ionization step, and membrane glycerophospholipids such as phosphatidylinositols (PIs), phosphatidylcholines (PCs), and phosphatidylserines (PSs). Main lipid classes detected in positive ion mode in brain tissue are PCs, phosphatidylethanolamine (PEs), sphingomyelins (SMs).

14. Reactive DESI experiment involves chemical derivatization performed simultaneously with ionization to generate a more favorable form of the analyte. Online derivatization has several advantages that are beneficial for ambient ionization, including (1) increased ionization efficiency and minimized ion suppression in complex biological matrices, and (2) enhanced chemical specificity to distinguish structural isomers or recover detailed structural information, e.g., double bond positions in lipids or peptide and protein characterization. The use of $AgNO_3$ in the DESI spray allows one to obtain lipid silver adducts whose ionization efficiency is greatly enhanced [28].

15. Using reactive DESI-MS with silver cationization, several cholesterol esters (CEs), TGs, DGs, ubiquinone, and squalene have been detected in human and mouse brain tissue, zebrafish, bovine, porcine, and mouse oocytes and embryos [12, 16]. Lipid silver adducts are easily recognized by the characteristic 1:1 abundance ratio of the ^{107}Ag:^{109}Ag isotopes. The adduction with silver confers a characteristic isotopic pattern that can be used to ease identification, even in low-resolution full-scan MS (Fig. S3b). The ions have been attributed mainly to the molecular species $[M+Ag]^+$ and $[M+Ag_2NO_3]^+$. For a few compounds, both molecular species can be simultaneously detected (e.g., ubiquinone), and their relative ratio in terms of ion intensity can be changed to favor either one species by

simply increasing ($[M+Ag_2NO_3]^+$>$[M+Ag]^+$) or decreasing ($[M+Ag]^+$>($[M+Ag_2NO_3]^+$) the concentration of silver nitrate in ACN.

16. Different solvent systems may require a different geometry of the DESI spray to have optimal spot size, signal intensity, and stability. Once the DESI spray is optimized for a specific solvent system, it is recommended to run all samples under the same experimental conditions to minimize analytical variability. It is common practice in our laboratory to build a DESI sprayer for each solvent system.

17. Negative and positive ion mode spectra are typically acquired using normal MS scan speeds; however, when working with ACN doped with silver nitrate, increased scan speeds are necessary to avoid space charge effects. The alternative scan types that improve mass resolution (e.g., enhanced scan) allow better deconvolution of the isotopic pattern of the silver $[M+Ag]^+$ and silver nitrate $[M+Ag_2NO_3]^+$ adducts and lead to overall higher signal-to-noise (Fig. S3).

18. Both DMF—ACN and pure ACN doped with $AgNO_3$ are histologically compatible solvents [11]. The term refers to spray solvents that do not cause morphological damage to the tissue. The mechanism behind DESI-MS is a spot-by-spot microextraction of molecules from the sample. The extraction process occurring with morphologically friendly solvents is comparable to the fixative procedures commonly used in histology for morphological examination: lipids are removed whereas the intracellular and extracellular structures stained in the histochemical treatment are preserved.

19. Different types of multi-block data acquired by DESI-MS analysis have been already presented in the literature:

(a) same solvent system (DMF—ACN, 1:1 v/v) used to acquire MS data over two mass-to-charge ranges (m/z 80–200 and m/z 700–1000), in order to detect small metabolites, such as lactic acid, ascorbic acid, N-acetyl-aspartic acid, and glycerophospholipids, respectively [22].

(b) same m/z range (700–1000) and solvent system (DMF—ACN, 1:1 v/v) used to acquire MS data in negative and positive ion modes to detect different classes of glycerophospholipids [46].

(c) different mass ranges (m/z 700–1000 and 600–1400, respectively in negative and positive ion modes) and different solvent systems (DMF–ACN, 1:1 v/v, and ACN doped with $AgNO_3$, respectively) in order to detect glycerophospholipids as negative ions and TGs, DGs, ubiquinone, squalene, and CEs as positive ions [28, 29].

20. Linear ion traps are examples of low-resolution mass analyzers (mass resolving power $m/\Delta m = 1000$) capable of MS^n fragmentation.

21. Product ion scan ($\bullet \rightarrow \bigcirc$) is an MS scan of one dimension in which a precursor ion is isolated in the ion trap, fragmented, and all products carrying the residual charge are subsequently detected. It provides detailed information on a single ion and the neutral molecule from which it is a surrogate [12].

22. Typical settings are: collision energy, CE = 0 a.u. for initial isolation, CE = 15–35 a.u. for fragmentation (note that a.u. refers to a manufacture unit but the CE values can be converted into eV); isolation width = m/z 1.7; Activation q = 0.25; activation time = 30 ms. Low-energy CID MS/MS has the advantage to be relatively reproducible in the fragments obtained for a specific compound; however, different instruments and different settings for the same instrument (e.g., CE value) can change the relative ion intensity between fragments.

23. Low-resolution DESI-MS/MS images of product ions detected for a precursor ion can reveal different spatial distributions of isobaric compounds within the sample. Indeed, isobaric (even isomeric) compounds can fragment following different pathways, thus producing selective products. Spatial differences between isobaric species cannot be appreciated when acquiring the nominal masses only in low-resolution full-scan DESI-MS images.

24. For example, low-energy CID is not effective in locating sites of unsaturation because of the high dissociation energy involved in cleaving a C=C bond or even an allylic C–C bond. Several approaches have been proposed to tackle this problem: (1) chemical derivatization that either cleaves the bond, as in ozonolysis, or converts C=C bond into functional groups (e.g., alkylthiolation and methoxymercuration) that can be more easily fragmented by low-energy CID; (2) photochemical Patterno'-Buchi reaction; (3) use of more energetic dissociation mechanisms, e.g., high-energy CID (keV), which can favor charge remote fragmentation, or radical directed dissociation and metastable-atom activation dissociation [35–39].

25. https://metlin.scripps.edu/metabo_search_alt2.php. Additional free online resources for searching lipid exact masses or candidate formulas are LIPID MAPS (http://www.lipid-maps.org/) and Human Metabolome Database (HMDB, http://www.hmdb.ca/).

26. The mass error expressed as ppm is calculated as (Measured mass—Monoisotopic Exact Mass)/Monoisotopic Exact Mass* 1e6. The free program IsoPro 3.1 (https://sites.google.com/

site/isoproms/) is recommended for calculating the monoisotopic masses. Simply, click on the element of interest and copy the monoisotopic mass.

27. Errors <5 ppm are usually considered acceptable for identification. However, the stringency applied by the ppm mass error tolerance is dictated by the performance of the instrument used to collect the data.

28. Be aware that the chemical formula of a compound of interest might simply not be listed in a database. A software might help in figuring out possible formulae based on the high-resolution mass measurements. See, for example, the online software Molecular Formula Calculator (https://nationalmaglab.org/user-facilities/icr/icr-software). Type the exact mass measurement of the target peak, select the possible charge state and the maximum mass error in ppm. The software will list possible chemical formulae. The computation can be supported by specifying to the software which atoms are likely to be present in the structure.

29. The software IsoPro 3.1 can be also used to simulate the isotopic distribution of candidate formulas. Click on each atom present in the candidate molecular formula and specify its number (e.g., 59 atoms of carbons), and then select "calculate" and "distribution" or "display" to see the peaks list or visualize the spectrum, respectively. Note that in the parameter window, one can change the resolving power, the peak shape (Gaussian vs. Lorentzian), the ionization, and the charge state. Also LIPID MAPS offers a tool to predict structure and isotopic distribution profiles for a lipid ion of interest. Choose the functional groups and the ion type from the menu options. The software will return the exact mass and the molecular formula as well (http://www.lipidmaps.org/tools/structure-drawing/masscalc.php).

30. In MS, isobaric species are atomic or molecular species with the same nominal mass but different exact masses.

31. Low-energy CID gives rise to predominantly charge-driven fragmentation processes, ideal for acyl chain and lipid class identification [35].

32. Product ion scan is the most efficient way of collecting MS/MS or MS^n (i.e., sequential product ion scan) data for the identification of target ions (one precursor ion by one) using ion traps. Neutral losses ($O \rightarrow O$) and precursor scans ($O \rightarrow \bullet$) [12] are very effective scans to simultaneously search for classes of lipids sharing common structural motifs in triple quadrupole MS systems. These scans can be indirectly reconstructed in time with ion traps by running a consecutive series of product scans and exploring the MS/MS data domain, as shown in Fig. S4.

However, the time necessary to collect such an MS/MS data domain can be as long as a few minutes, depending on the injection time set for each product scan and the range of precursor ions to be screened, and therefore is not ideal for a DESI-MS experiment based on a localized micro-extraction phenomenon.

Acknowledgments

CRF was supported from the Purdue University Center for Cancer Research Small Grants and from the Brazilian National Council for Scientific and Technological Development (CNPq). VP gladly acknowledges the American Society for Mass Spectrometry for providing financial support (ASMS Postdoctoral Award 2015). The funders had no role in study design, data collection and analysis, decision to publish, or preparation of the manuscript.

References

1. Cooks RG, Ouyang Z, Takats Z, Wiseman JM (2006) Detection technologies. Ambient mass spectrometry. Science 311:1566–1570. doi:10.1126/science.1119426

2. Badu-Tawiah AK, Eberlin LS, Ouyang Z, Cooks RG (2013) Chemical aspects of the extractive methods of ambient ionization mass spectrometry. Annu Rev Phys Chem 64:481–505. doi:10.1146/annurev-physchem-040412-110026

3. Takats Z, Wiseman JM, Gologan B, Cooks RG (2004) Mass spectrometry sampling under ambient conditions with desorption electrospray ionization. Science 306:471–473. doi:10.1126/science.1104404

4. Cooks RG, Jarmusch AK, Ferreira CR, Pirro V (2015) Skin molecule maps using mass spectrometry. Proc Natl Acad Sci U S A 112:5261–5262. doi:10.1073/pnas.1505313112

5. Monge ME, Harris GA, Dwivedi P, Fernandez FM (2013) Mass spectrometry: recent advances in direct open air surface sampling/ionization. Chem Rev 113:2269–2308. doi:10.1021/cr300309q

6. Eberlin LS (2014) DESI-MS imaging of lipids and metabolites from biological samples. Methods Mol Biol 1198:299–311. doi:10.1007/978-1-4939-1258-2_20

7. Santagata S, Eberlin LS, Norton I, Calligaris D, Feldman DR, Ide JL, Liu X, Wiley JS, Vestal ML, Ramkissoon SH, Orringer DA, Gill KK, Dunn IF, Dias-Santagata D, Ligon KL, Jolesz FA, Golby AJ, Cooks RG, Agar NY (2014) Intraoperative mass spectrometry mapping of an onco-metabolite to guide brain tumor surgery. Proc Natl Acad Sci U S A 111:11121–11126. doi:10.1073/pnas.1404724111

8. Siebenhaar M, Kullmer K, de Barros Fernandes NM, Hullen V, Hopf C (2015) Personalized monitoring of therapeutic salicylic acid in dried blood spots using a three-layer setup and desorption electrospray ionization mass spectrometry. Anal Bioanal Chem 407:7229–7238. doi:10.1007/s00216-015-8887-8

9. Talaty N, Mulligan CC, Justes DR, Jackson AU, Noll RJ, Cooks RG (2008) Fabric analysis by ambient mass spectrometry for explosives and drugs. Analyst 133:1532–1540. doi:10.1039/b807934j

10. Eberlin LS, Ferreira CR, Dill AL, Ifa DR, Cooks RG (2011) Desorption electrospray ionization mass spectrometry for lipid characterization and biological tissue imaging. Biochim Biophys Acta 1811(11)946–960. doi:10.1016/j.bbalip.2011.05.006

11. Eberlin LS, Ferreira CR, Dill AL, Ifa DR, Cheng L, Cooks RG (2011) Nondestructive, histologically compatible tissue imaging by desorption electrospray ionization mass spectrometry. Chembiochem 12:2129–2132. doi:10.1002/cbic.201100411

12. Ferreira CR, Jarmusch AK, Pirro V, Alfaro CM, Gonzalez-Serrano AF, Niemann H, Wheeler MB, Rabel RA, Hallett JE, Houser R,

Kaufman A, Cooks RG (2015) Ambient ionisation mass spectrometry for lipid profiling and structural analysis of mammalian oocytes, pre-implantation embryos and stem cells. Reprod Fertil Dev 27:621–637. doi:10.1071/RD14310

13. Ferreira CR, Pirro V, Jarmusch AK, Alfaro CM, Cooks RG (2016) Ambient lipidomic analysis of single mammalian oocytes and pre-implantation embryos using desorption electrospray ionization (DESI) mass spectrometry. Methods Mol Biol. Accepted for publication

14. Ferreira CR, Eberlin LS, Hallett JE, Cooks RG (2012) Single oocyte and single embryo lipid analysis by desorption electrospray ionization mass spectrometry. J Mass Spectrom 47:29–33. doi:10.1002/jms.2022

15. Eberlin LS, Liu X, Ferreira CR, Santagata S, Agar NY, Cooks RG (2011) Desorption electrospray ionization then MALDI mass spectrometry imaging of lipid and protein distributions in single tissue sections. Anal Chem 83:8366–8371. doi:10.1021/ac202016x

16. Pirro V, Guffey SC, Sepúlveda MS, Mahapatra CT, Ferreira CR, Jarmusch AK, Cooks RG (2016) Lipid dynamics in zebrafish embryonic development observed by DESI-MS imaging and nanoelectrospray-MS. Mol Biosyst 12:2069–2079. doi:10.1039/C6MB00168H

17. Kerian KS, Jarmusch AK, Pirro V, Koch MO, Masterson TA, Cheng L, Cooks RG (2015) Differentiation of prostate cancer from normal tissue in radical prostatectomy specimens by desorption electrospray ionization and touch spray ionization mass spectrometry. Analyst 140:1090–1098. doi:10.1039/c4an02039a

18. Dill AL, Eberlin LS, Costa AB, Zheng C, Ifa DR, Cheng L, Masterson TA, Koch MO, Vitek O, Cooks RG (2011) Multivariate statistical identification of human bladder carcinomas using ambient ionization imaging mass spectrometry. Chemistry 17:2897–2902. doi:10.1002/chem.201001692

19. Dill AL, Eberlin LS, Zheng C, Costa AB, Ifa DR, Cheng L, Masterson TA, Koch MO, Vitek O, Cooks RG (2010) Multivariate statistical differentiation of renal cell carcinomas based on lipidomic analysis by ambient ionization imaging mass spectrometry. Anal Bioanal Chem 398:2969–2978. doi:10.1007/s00216-010-4259-6

20. Eberlin LS, Tibshirani RJ, Zhang J, Longacre TA, Berry GJ, Bingham DB, Norton JA, Zare RN, Poultsides GA (2014) Molecular assessment of surgical-resection margins of gastric cancer by mass-spectrometric imaging. Proc Natl Acad Sci U S A 111:2436–2441. doi:10.1073/pnas.1400274111

21. Jarmusch AK, Kerian KS, Pirro V, Peat T, Thompson CA, Ramos-Vara JA, Childress MO, Cooks RG (2015) Characteristic lipid profiles of canine non-Hodgkin's lymphoma from surgical biopsy tissue sections and fine needle aspirate smears by desorption electrospray ionization—mass spectrometry. Analyst 140:6321–6329. doi:10.1039/c5an00825e

22. Jarmusch AK, Pirro V, Baird Z, Hattab EM, Cohen-Gadol AA, Cooks RG (2016) Lipid and metabolite profiles of human brain tumors by desorption electrospray ionization-MS. Proc Natl Acad Sci U S A 113:1486–1491. doi:10.1073/pnas.1523306113

23. Eberlin LS, Norton I, Dill AL, Golby AJ, Ligon KL, Santagata S, Cooks RG, Agar NY (2012) Classifying human brain tumors by lipid imaging with mass spectrometry. Cancer Res 72:645–654. doi:10.1158/0008-5472.CAN-11-2465

24. Eberlin LS, Norton I, Orringer D, Dunn IF, Liu X, Ide JL, Jarmusch AK, Ligon KL, Jolesz FA, Golby AJ, Santagata S, Agar NY, Cooks RG (2013) Ambient mass spectrometry for the intraoperative molecular diagnosis of human brain tumors. Proc Natl Acad Sci U S A 110:1611–1616. doi:10.1073/pnas.1215687110

25. Righi V, Roda JM, Paz J, Mucci A, Tugnoli V, Rodriguez-Tarduchy G, Barrios L, Schenetti L, Cerdan S, Garcia-Martin ML (2009) 1H HR-MAS and genomic analysis of human tumor biopsies discriminate between high and low grade astrocytomas. NMR Biomed 22:629–637. doi:10.1002/nbm.1377

26. Jain M, Nilsson R, Sharma S, Madhusudhan N, Kitami T, Souza AL, Kafri R, Kirschner MW, Clish CB, Mootha VK (2012) Metabolite profiling identifies a key role for glycine in rapid cancer cell proliferation. Science 336:1040–1044. doi:10.1126/science.1218595

27. Beloribi-Djefaflia S, Vasseur S, Guillaumond F (2016) Lipid metabolic reprogramming in cancer cells. Oncogenesis 5:e189. doi:10.1038/oncsis.2015.49

28. Gonzalez-Serrano AF, Pirro V, Ferreira CR, Oliveri P, Eberlin LS, Heinzmann J, Lucas-Hahn A, Niemann H, Cooks RG (2013) Desorption electrospray ionization mass spectrometry reveals lipid metabolism of individual oocytes and embryos. PLoS One 8:e74981. doi:10.1371/journal.pone.0074981

29. Pirro V, Oliveri P, Ferreira CR, Gonzalez-Serrano AF, Machaty Z, Cooks RG (2014) Lipid characterization of individual porcine

oocytes by dual mode DESI-MS and data fusion. Anal Chim Acta 848:51–60. doi:10.1016/j.aca.2014.08.001

30. Robichaud G, Garrard KP, Barry JA, Muddiman DC (2013) MSiReader: an open-source interface to view and analyze high resolving power MS imaging files on Matlab platform. J Am Soc Mass Spectrom 24:718–721. doi:10.1007/s13361-013-0607-z

31. Pirro V, Eberlin LS, Oliveri P, Cooks RG (2012) Interactive hyperspectral approach for exploring and interpreting DESI-MS images of cancerous and normal tissue sections. Analyst 137:2374–2380. doi:10.1039/c2an35122f

32. Fahy E, Subramaniam S, Brown HA, Glass CK, Merrill AH Jr, Murphy RC, Raetz CR, Russell DW, Seyama Y, Shaw W, Shimizu T, Spener F, van Meer G, VanNieuwenhze MS, White SH, Witztum JL, Dennis EA (2005) A comprehensive classification system for lipids. J Lipid Res 46:839–861. doi:10.1194/jlr.E400004-JLR200

33. Fahy E, Subramaniam S, Murphy RC, Nishijima M, Raetz CR, Shimizu T, Spener F, van Meer G, Wakelam MJ, Dennis EA (2009) Update of the LIPID MAPS comprehensive classification system for lipids. J Lipid Res 50:S9–14. doi:10.1194/jlr.R800095-JLR200

34. Liebisch G, Vizcaíno JA, Köfeler H, Trötzmüller M, Griffiths WJ, Schmitz G, Spener F, Wakelam MJ (2013) Shorthand notation for lipid structures derived from mass spectrometry. J Lipid Res 54:1523–1530. doi:10.1194/jlr.M033506

35. Murphy RC (2015) Tandem mass spectrometry of lipids: molecular analysis of complex lipids. Royal Society of Chemistry, Cambridge, p 280

36. Ma X, Chong L, Tian R, Shi R, Hu TY, Ouyang Z, Xia Y (2016) Identification and quantitation of lipid C=C location isomers: a shotgun lipidomics approach enabled by photochemical reaction. Proc Natl Acad Sci U S A 113:2573–2578. doi:10.1073/pnas.1523356113

37. Blanksby SJ, Mitchell TW (2010) Advances in mass spectrometry for lipidomics. Annu Rev Anal Chem 3:433–465. doi:10.1146/annurev.anchem.111808.073705

38. Kozlowski RL, Mitchell TW, Blanksby SJ (2015) A rapid ambient ionization-mass spectrometry approach to monitoring the relative abundance of isomeric glycerophospholipids. Sci Rep 5:9243. doi:10.1038/srep09243

39. Pham HT, Maccarone AT, Thomas MC, Campbell JL, Mitchell TW, Blanksby SJ (2014) Structural characterization of glycerophospholipids by combinations of ozone- and collision-induced dissociation mass spectrometry: the next step towards "top-down" lipidomics. Analyst 139:204–214. doi:10.1039/c3an01712e

40. Manicke NE, Dill AL, Ifa DR, Cooks RG (2010) High resolution tissue imaging on an orbitrap mass spectrometer by desorption electro-spray ionization mass spectrometry (DESI-MS) J Mass Spectrom 45:223–226. doi:10.1002/jms.1707

41. Brugger B, Erben G, Sandhoff R, Wieland FT, Lehmann WD (1997) Quantitative analysis of biological membrane lipids at the low picomole level by nano-electrospray ionization tandem mass spectrometry. Proc Natl Acad Sci U S A 94:2339–2344

42. Schneiter R, Brugger B, Sandhoff R, Zellnig G, Leber A, Lampl M, Athenstaedt K, Hrastnik C, Eder S, Daum G, Paltauf F, Wieland FT, Kohlwein SD (1999) Electrospray ionization tandem mass spectrometry (ESI-MS/MS) analysis of the lipid molecular species composition of yeast subcellular membranes reveals acyl chain-based sorting/remodeling of distinct molecular species en route to the plasma membrane. J Cell Biol 146:741–754

43. Roscioli KM, Tufariello JA, Zhang X, Li SX, Goetz GH, Cheng G, Siems WF, Hill HH Jr (2014) Desorption electrospray ionization (DESI) with atmospheric pressure ion mobility spectrometry for drug detection. Analyst 139:1740–1750. doi:10.1039/c3an02113k

44. Myung S, Wiseman JM, Valentine SJ, Takats Z, Cooks RG, Clemmer DE (2006) Coupling desorption electrospray ionization with ion mobility/mass spectrometry for analysis of protein structure: evidence for desorption of folded and denatured States. J Phys Chem B 110:5045–5051. doi:10.1021/jp052663e

45. Škrášková K, Claude E, Jones EA, Towers M, Ellis SR, Heeren RM (2016) Enhanced capabilities for imaging gangliosides in murine brain with matrix-assisted laser desorption/ionization and desorption electrospray ionization mass spectrometry coupled to ion mobility separation. Methods S1046-2023:30031–30037. doi:10.1016/j.ymeth.2016.02.01446.

46. Jarmusch AK, Alfaro CM, Pirro V, Hattab EM, Cohen-Gadol AA, Cooks RG (2016) Differential Lipid Profiles of Normal Human Brain Matter and Gliomas by Positive and Negative Mode Desorption Electrospray Ionization - Mass Spectrometry Imaging. PLoS One 11: e0163180. doi:10.1371/journal.pone.0163180.

Chapter 15

Lipidomics Analyses of Oxygenated Metabolites of Polyunsaturated Fatty Acids

Alexandra C. Kendall and Anna Nicolaou

Abstract

Prostanoids and hydroxy fatty acids derived from polyunsaturated fatty acids are widely recognized as mediators of inflammatory and homeostatic processes in the body. Accurate identification and quantification of these lipid mediators can aid in understanding of lipid metabolism in different tissues under different physiological and pathological conditions, potentially leading to biomarker discovery and identification of therapeutic targets, as well as providing support for the efficacy of novel therapies. The protocol outlined here describes the extraction of these lipids from a range of biological samples (including solid tissues, biological fluids, and cell culture media), their semi-purification by solid-phase extraction, and their analysis by reverse-phase ultra-performance liquid chromatography coupled to electrospray ionization tandem mass spectrometry. This provides highly sensitive and specific quantitative analysis of bioactive lipid species.

Key words Solid-phase extraction, Ultra-performance liquid chromatography, Electrospray ionization, Triple quadrupole mass spectrometry, Eicosanoids

1 Introduction

Oxygenated metabolites of polyunsaturated fatty acids (PUFA) can act as potent mediators of inflammation, as well as performing crucial roles in tissue homeostasis [1, 2]. Metabolism of the 20-carbon PUFA arachidonic acid (AA; 20:4n-6), eicosapentaenoic acid (EPA; 20:5n-3), and dihomo-γ-linolenic acid (DGLA; 20:3n-6) by cyclooxygenase (COX) enzymes produces prostanoids—members of the eicosanoid family that includes prostaglandins, thromboxanes, and prostacyclins (Fig. 1). These 20-carbon PUFA can also be metabolized by lipoxygenase (LOX) or cytochrome P450 (CYP) enzymes, to generate a range of hydroxy- and epoxy- fatty acid eicosanoid species. Additionally, LOX or CYP metabolism of the 18-carbon linoleic acid (LA) and α-linolenic acid (ALA; 18:3n-3), or the 22-carbon docosahexaenoic acid (DHA; 22:6n-3) generates octadecanoids or docosanoids, respectively [3]. Additionally, transcellular metabolism allows for

Paul Wood (ed.), *Lipidomics*, Neuromethods, vol. 125,
DOI 10.1007/978-1-4939-6946-3_15, © Springer Science+Business Media LLC 2017

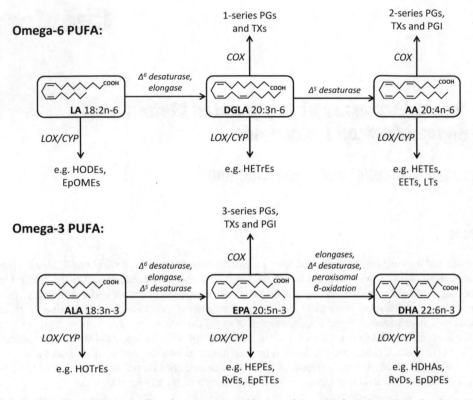

Fig. 1 Schematic outline of the production of prostanoids and hydroxy fatty acids from the omega-6 polyunsaturated fatty acids (PUFA) linoleic acid (LA), dihomo-γ-linolenic acid (DGLA) and arachidonic acid (AA), and the omega-3 PUFA α-linolenic acid, eicosapentaenoic acid (EPA) and docosahexaenoic acid (DHA) by cyclooxygenase (COX), lipoxygenase (LOX), and cytochrome P450 (CYP) enzymes

the step-wise synthesis of further lipid mediators, starting with LOX-metabolism by one cell type, then transport to another cell type for metabolism by a different LOX enzyme. This generates complex lipid mediators such as lipoxins and resolvins [4].

While enzyme-linked immunosorbent assays (ELISAs) exist for individual lipid species, these are limited by antibody availability and cross-reactivity [5]. Gas-chromatography (GC) linked to mass spectrometry (MS) is another option, although the need to derivatize compounds makes it unsuitable for some of the more thermally labile lipid species [6]. Because of these limitations, analysis of bioactive lipid mediators by liquid chromatography coupled to tandem mass spectrometry with electrospray ionization (LC/ESI-MS/MS) is the gold standard in accurate quantification [7]. The technology has high sensitivity and specificity, and the wide availability of commercially available standards, as well as deuterated internal standards, allows for accurate quantification. Thus, LC/ESI-MS/MS has become a potent tool to measure the in vitro, ex vivo, or in vivo production of bioactive lipid mediators from PUFA [8].

The assay we describe here allows for the extraction, analysis, and quantification of 79 prostanoids and hydroxy fatty acids from a range of biological fluids and tissues. The combined extraction of prostanoids and hydroxy fatty acids in an acidified methanol solution, followed by semi-purification by solid-phase extraction (SPE), produces a cleaner biological extract, with reduced matrix effects that may otherwise cause ion suppression [9, 10]. We use ultra-performance liquid chromatography (UPLC), which allows for high-throughput analysis, and we separate the analysis into two different LC/ESI-MS/MS assays, totaling just 10.8 min per sample. By analyzing the prostanoids and hydroxy fatty acids separately, we reduce co-elution of compounds and therefore improve assay specificity. ESI is the preferred ionization method for eicosanoids and related species, forming negative ion species (M-H$^-$) immediately post-elution from the column, reducing the risk of thermal decomposition [8]. Using commercially available standards of each compound, we have identified unique fragmentation patterns and/or retention times, which permit accurate identification of compounds by multiple reaction monitoring (MRM), including isobaric or co-eluting species [9, 10]. We use deuterated internal standards for normalization between injections, and produce calibration lines using the commercially available standards for accurate quantification.

2 Materials

Prepare all solutions using ultrapure water (resistivity of 18 MΩ cm at 25 °C). Analytical grade reagents should be used for lipid extraction. UPLC grade reagents should be used for liquid chromatography. Prepare and store all reagents at room temperature. Waste materials including solvents and biological samples should be disposed of according to local regulations.

2.1 Equipment

1. 1 ml Dounce homogenizer with tight-fitting pestle (Wheaton; Fisher Scientific, Loughborough, UK).

2. Blade homogenizer with 10 mm diameter shaft (Ystral X10/25; Scientific Instrument Centre, Winchester, UK).

3. Glass syringes (10 μl, 50 μl, 100 μl, 250 μl, 500 μl).

4. SPE vacuum manifold (Phenomenex).

5. Solvent drying cabinet with nitrogen supply.

6. Amber glass vials (2 ml) with 200 μl conical glass inserts and 8 mm screw caps with PTFE septa.

7. Sealing film (Parafilm®; Fisher Scientific, Loughborough, UK).

8. Waters Acquity UPLC pump with autosampler coupled to ESI triple quadrupole TQ-S mass spectrometer (Waters, Wilmslow, UK).

9. C18 column (Acquity UPLC BEH, 1.7 μm, 2.1 × 50 mm; Waters, Wilmslow, UK).

2.2 Reagents and Supplies

2.2.1 Standard Solutions for Quantification

1. The following standards for the prostanoid assay were obtained from Cayman Chemicals (Ann Arbor, MI, USA): prostaglandin E_1 (PGE$_1$), PGD$_1$, 6-keto PGF$_{1\alpha}$, 13,14-dihydro-15-keto PGE$_1$, 13,14-dihydro-15-keto PGF$_{1\alpha}$, PGF$_{1\alpha}$, PGD$_2$, PGE$_2$, 15-keto PGE$_2$, 13,14-dihydro PGE$_1$, 13,14-dihydro PGF$_{2\alpha}$, 13,14-dihydro PGF$_{1\alpha}$, 13,14-dihydro-15-keto PGF$_{2\alpha}$, PGF$_{2\alpha}$, 8-*isoprostane* PGF$_{2\alpha}$, PGF$_{3\alpha}$, Δ^{12}-PGJ$_2$, PGJ$_2$, thromboxane B$_2$ (TXB$_2$), TXB$_3$, 15-deoxy-$\Delta^{12,14}$-PGJ$_2$, 13,14-dihydro-15-keto PGE$_2$, PGD$_3$, PGE$_3$.

2. The following standards for the hydroxy fatty acid assay were obtained from Cayman Chemicals (Ann Arbor, MI, USA): 9-hydroxyoctadecadienoic acid (HODE) 13-HODE, resolvin D$_1$ (RvD$_1$), RvD$_2$, maresin$_1$ (MaR)$_1$, protectin DX (PDX), 11-hydroxydocosahexaenoic acid (11-HDHA), 4-HDHA, 7-HDHA, 10-HDHA, 13-HDHA, 14-HDHA, 17-HDHA, 20-HDHA, RvE$_1$, leukotriene B$_4$ (LTB$_4$), 14,15 dihydroxyeicosatrienoic acid (14,15-DHET), 11,12-DHET, 8,9-DHET, 5,6-DHET, 5(6) epoxyeicosatrienoic acid (5(6)-EET), 11(12)-EET, 14(15)-EET, 8(9)-EET, 5-oxo-eicosatetraenoic acid (5-oxo-ETE), 15-hydroxyeicosatrienoic acid (15-HETrE), 5 hydroxyeicosatetraenoic acid (5-HETE), 8-HETE, 9-HETE, 11-HETE, 12-HETE, 15-HETE, 20-HETE, 5 hydroxyeicosapentaenoic acid (5-HEPE), 8-HEPE, 9-HEPE, 11-HEPE, 15-HEPE, 18-HEPE, 12-HEPE, 9 hydroxyoctadecatrienoic acid (9 HOTrE), 13 HOTrE, 19(20) dihydroxydocosapentaenoic acid (DiHDPA), 9(10) epoxyoctadecenoic acid (EpOME), 12(13) EpOME, 9(10) dihydroxyoctadecenoic acid (DiHOME), 12(13) DiHOME, 9 oxooctadecadienoic acid (OxoODE), 13 OxoODE, 5(15) dihydroxyeicosatetraenoic acid (DiHETE), 8(15) DiHETE, 19(20) epoxydocosapentaenoic acid (EpDPE), 16(17) EpDPE, trans epoxyketooctadecenoic acid (EKODE).

3. Ethanol (HPLC grade; Fisher Scientific, Loughborough, UK).

4. Cryogenic storage boxes for 2 ml glass vials.

2.2.2 Internal Standard Cocktail

1. The following deuterated standards were obtained from Cayman Chemicals (Ann Arbor, MI, USA): PGB$_2$-*d4*, 12-HETE-*d8*, 8(9)EET-*d11*, 8,9-DHET-*d11*.

2. Ethanol (HPLC grade; Fisher Scientific, Loughborough, UK).

3. Cryogenic storage boxes for 2 ml glass vials.

2.2.3 Biological Sample Preparation

1. Dissecting equipment (tweezers, scissors).
2. Glass Pasteur pipettes.
3. Flat-bottomed 10 ml wide-neck glass extraction tubes with lids.
4. Round-bottomed 10 ml glass extraction tubes with lids.
5. pH indicator strips.
6. 15% methanol solution.
7. HCl (0.1 M or 1 M).

2.2.4 Solid-Phase Extraction

1. Methanol, hexane, methyl formate (Sigma-Aldrich, Poole, UK).
2. STRATA™ SPE cartridges (C18-E, 500 mg, 6 ml; Phenomenex, Macclesfield, UK).

2.2.5 UPLC/ESI-MS/MS

1. Mobile phase A: Water with 0.02% acetic acid. Measure 1 l water into a clean 1 l Duran bottle. Remove 200 μl water using a pipette and discard. Add 200 μl acetic acid to the bottle. Mix thoroughly (*see* **Note 1**).
2. Mobile phase B: acetonitrile with 0.02% acetic acid. Take a 2.5 l bottle of UPLC-grade acetonitrile. Remove 500 μl acetonitrile from the bottle using a pipette and discard. Add 500 μl acetic acid to the bottle. Mix thoroughly (*see* **Note 2**).
3. Seal wash: methanol:water (50:50 v/v). Measure 500 ml methanol in a 1 l measuring cylinder. Make up to 1 l with water. Transfer to a 1 l Duran bottle and mix thoroughly (*see* **Note 3**).
4. Shutdown solvent: 100% acetonitrile (*see* **Note 4**).

3 Methods

3.1 Standard Solutions

3.1.1 Stock and Working Solutions (Preparation and Storage)

1. Lipid standards as purchased are provided at various concentrations in solution in an organic solvent (usually methyl acetate or ethanol).
2. Working solutions of 10 ng/μl should be prepared from each lipid standard by adding the appropriate volume of stock solution to a 2 ml amber vial, and making the volume up to 1 ml with ethanol. These will be used to construct the calibration cocktail.
3. Both stock and working solutions should be flashed with nitrogen and sealed with sealing film, before storage at −80 °C (*see* **Note 5**).

3.1.2 Internal Standard Cocktail

1. 10 ng/μl working solutions of each internal standard should be prepared as in Sect. 3.1.1.
2. To make a 1 ng/μl internal standard cocktail, add 100 μl of each internal standard to a clean 2 ml amber glass vial.

3. Make up to 1 ml with ethanol and seal with an 8 mm screw cap and sealing film.

4. Store at −20 °C for up to 3 months.

3.1.3 Calibration Lines for Quantification

1. A 100 pg/μl standard cocktail for prostanoids and a separate 100 pg/μl standard cocktail for hydroxy fatty acids are made from the appropriate 10 ng/μl working solutions.

2. 10 μl of each standard (working solution 10 ng/μl) is transferred to a clean 2 ml amber vial using a glass syringe.

3. The solution is made up to 1 ml with ethanol, resulting in a final concentration of 100 pg/μl of each standard. The cocktail can be stored at −80 °C for up to 6 months.

4. Calibration lines for the prostanoid assay and the hydroxy fatty acid assay are made in the same way, but using the two different 100 pg/μl standard cocktails. The calibration lines are prepared by series dilution to give concentrations of 10 pg/μl, 5 pg/μl, 2.5 pg/μl, 1.25 pg/μl, 0.625 pg/μl, 0.313 pg/μl, and 0.156 pg/μl. Label 2 ml amber vials appropriately and add a 200 μl conical glass insert to each vial (*see* **Note 6**).

5. Add 180 μl ethanol to the 10 pg/μl vial, add 100 μl to all other vials. Take 20 μl of the appropriate standard cocktail to the 10 pg/μl vial, mix with the ethanol (using the glass syringe), and then transfer 100 μl to the 5 pg/μl vial and mix with the ethanol. Continue transferring 100 μl and mixing down the dilution series until the lowest concentration. Discard 100 μl from this vial after mixing, so that each vial contains 100 μl.

6. Add 20 μl of the internal standard cocktail (see Sect. 3.1.2) to each vial (so that each vial now contains 120 μl).

7. Dry down the contents of each vial under a stream of nitrogen.

8. Add 100 μl ethanol to each vial and mix thoroughly. Seal the vials with an 8 mm screw cap with a PTFE insert.

9. Calibration lines can be stored for up to a week at −20 °C, and a fresh calibration line should be prepared for each experiment.

3.2 Collection and Storage of Biological Samples

In principle, process samples as quickly as possible and on ice. Collect samples on ice, aliquot and freeze as soon as possible. Do not add any preservatives. Store at −80 °C, avoiding freeze/thaw cycles. If transport is necessary, this should be done on dry ice.

3.2.1 Plasma and Serum

1. Plasma (prepared from EDTA tubes) or serum should be aliquoted (500 μl) in cryogenic vials, and stored at −80 °C. This avoids freeze/thaw cycles, and if larger volumes are needed it is better to pool small aliquots than risk having to freeze/thaw larger ones.

3.2.2 Cell Culture Media

1. Cell culture media is collected from a known number of cells; the required number of cells will depend on their ability to produce lipid mediators. Typically, cells are grown in T25 or T75 flasks, and when they reach 80% confluency, the culture media is removed and replaced with 5 or 10 ml (T25 and T75, respectively) fresh media for the cells to undergo treatment. At the end of the treatment, the media is collected in 15 ml polypropylene vials and stored at −80 °C (*see* **Note 7**). Cells should be counted to give an estimate of the number of cells that contributed lipids to the media. Alternatively, lipid concentrations can be normalized against the protein content of cells.

3.2.3 Solid Tissue Samples

1. Tissue samples (20–200 mg) should be snap-frozen as soon as possible after collection to avoid enzymatic lipid degradation post-harvesting. Samples should then be stored at −80 °C.

3.3 Biological Sample Preparation

Solvent extraction should be performed on ice, keeping the samples out of direct light. Solutions should be prechilled on ice. We use glass tubes/pipettes/syringes wherever possible to avoid leaching of plasticizers into the solvent, and to minimize adhesion of the lipids to the tubes.

3.3.1 Plasma or Serum Samples

1. Defrost the samples on ice.
2. Transfer the sample to a flat-bottomed glass extraction tube by pipette, measuring and taking note of the sample volume (*see* **Note 8**).
3. Dilute the sample to a volume of 4 ml with water.
4. Add 700 μl methanol to make a 15% (v/v) methanol solution.
5. Add internal standard cocktail (20 μl of the prepared 1 ng/μl cocktail prepared in Sect. 3.1.2).
6. Mix the solution gently and leave on ice, in the dark for 15 min.
7. Centrifuge (5 min, 4 °C, 1500 × g).
8. Transfer the clear supernatant to a clean, prechilled, glass tube.
9. Acidify the supernatant to pH 1.7 using 0.1 M HCl (*see* **Note 9**).
10. Immediately apply the acidified sample to the conditioned SPE cartridge (*see* Sect. 3.4).

3.3.2 Cell Culture Media

1. Defrost the sample on ice.
2. Transfer the sample to a flat-bottomed glass extraction tube by pipette, measuring and taking note of the sample volume.
3. Add sufficient methanol to make a 15% methanol solution (e.g., 10 ml media will require the addition of 1.77 ml methanol).

4. Add internal standard cocktail (20 μl of the prepared 1 ng/μl cocktail prepared in Sect. 3.1.2).

5. Mix the solution gently and leave on ice, in the dark for 15 min.

6. Centrifuge (5 min, 4 °C, 1500 × *g*).

7. Transfer the clear supernatant to a clean, prechilled, glass tube (*see* **Note 10**).

8. Acidify the supernatant to pH 1.7 using 1 M HCl (*see* **Note 11**).

9. Immediately apply the acidified sample to the conditioned SPE cartridge (*see* Sect. 3.4).

3.3.3 Soft Tissue Samples (e.g., Brain, Kidney, Liver)

1. Defrost sample on ice (*see* **Note 12**).

2. Transfer sample to ice-cold mortar of a Dounce homogenizer using tweezers.

3. Add 1 ml ice-cold 15% methanol to the mortar and, using the tight-fitting pestle, perform 15 up/down strokes and 15 rotational strokes to homogenize the sample.

4. Using a glass Pasteur pipette, transfer the sample to a clean flat-bottomed glass vial.

5. Repeat steps 3–4 twice more, so that a total of 3 ml has been used and transferred to the glass tube (*see* **Note 13**).

6. Add internal standard cocktail (20 μl of the prepared 1 ng/μl cocktail prepared in Sect. 3.1.2).

7. Mix the solution gently and leave on ice, in the dark for 30 min (*see* **Note 14**).

8. Centrifuge (5 min, 4 °C, 1500 × *g*).

9. Transfer the clear supernatant to a clean, prechilled, glass tube and retain the pellet for protein content analysis (*see* **Note 15**).

10. Acidify the supernatant to pH 1.7 using 0.1 M HCl.

11. Immediately apply the acidified sample to the conditioned SPE cartridge (*see* Sect. 3.4).

3.3.4 Tough Tissue Samples (e.g., Skin, Muscle, Cornea)

1. Defrost sample on ice (*see* **Note 12**).

2. Mince tissue using dissection scissors into a flat-bottomed extraction tube containing 3 ml ice-cold 15% methanol.

3. Homogenize tissue using a blade homogenizer inserted into the tube, using 3 × 3 s pulses (*see* **Notes 16** and **13**).

4. Add internal standard cocktail (20 μl of the prepared 1 ng/μl cocktail prepared in Sect. 3.1.2).

5. Mix the solution gently and leave on ice, in the dark for 60 min (*see* **Note 14**).

6. Centrifuge (5 min, 4 °C, 1500 × *g*).

7. Transfer the clear supernatant to a clean, prechilled, glass tube and retain the pellet for protein content analysis (*see* **Note 15**).

8. Acidify the supernatant to pH 1.7 using 0.1 M HCl.

9. Immediately apply the acidified sample to the conditioned SPE cartridge (*see* Sect. 3.4).

3.4 Solid-Phase Extraction

3.4.1 SPE Cartridge Conditioning

1. Label and attach the required number of SPE cartridges to the SPE vacuum manifold, making sure any unused ports are closed and there is a waste bucket underneath the cartridges for waste solvent collection.

2. Activate the cartridges with 6 ml 100% methanol, followed by 6 ml water, under a gentle vacuum (approximately 3 mmHg) (*see* **Note 17**).

3. Close the tap under each cartridge as soon as the last of the water has flowed through (*see* **Note 18**).

3.4.2 SPE of Lipid Extracts

1. Transfer the acidified samples onto the conditioned cartridges using glass Pasteur pipettes, opening the tap under each cartridge as it is filled.

2. Let the sample flow through the SPE cartridges in a dropwise manner with the vacuum switched off (*see* **Note 19**).

3. Take care not to allow any air to pass through the cartridge sorbent and close the tap under each cartridge as soon as the last of the sample has flowed through.

4. Wash all the cartridges sequentially with 15% methanol solution (6 ml), water (6 ml), and hexane (6 ml) under a low vacuum, letting the eluate run into the waste and closing the tap under each cartridge as the last of the solvent runs through in each stage (*see* **Notes 20** and **21**).

5. Turn off the vacuum and remove the waste bucket from the vacuum manifold, replacing it with clean, round-bottomed glass tubes positioned under each SPE cartridge in a rack (*see* **Note 22**).

6. Elute the lipids with methyl formate (6 ml) (*see* **Note 23**).

7. Remove the tubes from the vacuum manifold and transfer to the drying apparatus.

3.5 Sample Reconstitution and Preparation for UPLC/ESI-MS/MS Analysis

1. Evaporate the methyl formate from the lipid extracts under a gentle stream of nitrogen (*see* **Note 24**).

2. Add 100 μl ethanol to each sample tube, vortex to ensure the entire inner surface is coated, and centrifuge briefly (10 s, $3000 \times g$, 4 °C).

3. Using a glass syringe, transfer the lipid extract to a 200 μl conical glass insert inside a 2 ml amber vial. Cap with an 8 mm screw cap with a PTFE septum. Seal with sealing film and store at −20 °C for analysis within 1 week.

3.6 UPLC/ESI-MS/MS

1. Condition the column by running UPLC solvents A and B at a ratio of 50:50 and at a flow rate of 0.6 ml/min for at least 1 h. At the same time, put the MS into operate mode for at least 1 h. Check that there is a stable back pressure on the UPLC and clean baseline on the MS.

2. Set the autosampler to 8 °C and add sample vials to the autosampler plates, setting up a sample list with appropriate vial positions.

3. Program the gradient profiles (MassLynx™ LC Inlet method file) as in Table 1 (prostanoid assay) and Table 2 (hydroxy fatty acid assay).

4. Program the MS Method files (MassLynx™ method page) using the MRM, cone voltage, and collision energy settings listed in Tables 3 and 4. Ensure the run times match the run times in the appropriate LC method. Set the dwell time to auto dwell.

5. Set the MassLynx™ Tune page for each assay with the settings below in Table 5.

6. Check that the API and collision gases are switched on.

Table 1
LC gradient for prostanoid assay (flow rate 0.6 ml/min)

Time (min)	Solvent A (%)	Solvent B (%)
0.0	80	20
0.5	80	20
0.6	60	40
2.5	60	40
4.0	35	65
4.1	80	20
5.8	80	20

Table 2
LC gradient for hydroxy fatty acid assay (flow rate 0.6 ml/min)

Time (min)	Solvent A (%)	Solvent B (%)
0.0	75	25
3.0	20	80
3.2	75	25
5.0	75	25

Table 3
Multiple reaction monitoring (MRM) transitions for prostanoid assay

Compound	MRM	Cone voltage (V)	Collision energy (eV)	Indicative retention time (min)
PGD$_1$	353 > 317	12	12	1.33
PGE$_1$	353 > 317	12	12	1.28
6-keto PGF$_{1\alpha}$	369 > 163	12	24	0.99
13,14-dihydro-15-keto PGF$_{1\alpha}$	355 > 193	20	30	1.62
PGF$_{1\alpha}$	355 > 311	14	24	1.17
13,14-dihydro PGE$_1$	355 > 337	18	16	1.40
13,14-dihydro-15-keto PGE$_1$	353 > 335	12	14	1.69
PGB$_2$-$d4$	337 > 179	12	20	2.06
PGD$_2$	351 > 271	24	16	134
PGE$_2$	351 > 271	24	16	1.25
15-keto PGE$_2$	349 > 113	14	20	1.40
13,14-dihydro PGF$_{2\alpha}$	355 > 311	14	24	1.32
13,14-dihydro-15-keto PGF$_{2\alpha}$	353 > 113	10	26	1.69
PGF$_{2\alpha}$	353 > 193	12	24	1.18
8-iso PGF$_{2\alpha}$	353 > 193	12	24	1.10
PGJ$_2$	333 > 271	14	16	1.99
Δ^{12}-PGJ$_2$	333 > 271	14	16	2.03
15-deoxy-$\Delta^{12,14}$-PGJ$_2$	315 > 271	12	14	3.90
TXB$_2$	369 > 169	18	18	1.11
13,14-dihydro PGF$_{1\alpha}$	357 > 113	4	32	1.37
13,14-dihydro-15-keto PGE$_2$	351 > 333	12	12	1.57
TXB$_3$	367 > 169	16	14	1.03
PGD$_3$	349 > 269	10	16	1.18
PGE$_3$	349 > 269	10	16	1.13
PGF$_{3\alpha}$	351 > 193	2	22	1.08

7. Load the LC Inlet method for the prostanoid assay—the solvent ratio and flow rate should change to the starting conditions of the prostanoid LC method.

8. Program the sample list with the appropriate MS Method, LC Inlet, and Tune files. Give each standard and sample a 3 µl injection volume (*see* **Note 26**).

Table 4
Multiple reaction monitoring (MRM)transitions for hydroxy fatty acid assay (*see* Note 25)

Compound	MRM	Cone voltage (V)	Collision energy (eV)	Indicative retention time (min)
9-HODE	295 > 171	16	16	2.47
13-HODE	295 > 195	2	18	2.49
15-HETrE	321 > 303	2	14	2.68
5-HETE	319 > 115	14	14	2.75
8-HETE	319 > 155	10	14	2.66
9-HETE	319 > 123	16	14	2.71
11-HETE	319 > 167	14	14	2.61
12-HETE	319 > 179	20	14	2.66
12-HETE-*d*8	327 > 184	20	16	2.64
15-HETE	319 > 175	4	14	2.54
20 HETE	319 > 245	4	14	2.31
5(6)-EET	319 > 191	4	10	3.03
8(9)-EET	319 > 155	10	14	2.66
11(12)-EET	319 > 167	14	14	2.61
14(15)-EET	319 > 175	4	14	2.54
5-HEPE	317 > 115	16	12	2.46
8-HEPE	317 > 155	26	12	2.38
9-HEPE	317 > 149	20	14	2.42
11-HEPE	317 > 167	12	12	2.35
12-HEPE	317 > 179	28	12	2.39
15-HEPE	317 > 175	8	14	2.33
18-HEPE	317 > 215	12	14	2.24
5,6-DHET	337 > 145	8	16	2.33
8,9-DHET	337 > 127	8	16	2.22
11,12-DHET	337 > 167	2	18	2.14
14, 15-DHET	337 > 207	18	16	2.04
5-oxo-ETE	317 > 203	14	18	2.94
LTB$_4$	335 > 195	12	14	1.87
RvE$_1$	349 > 195	14	16	0.81
RvD$_1$	375 > 141	18	12	1.39

(continued)

Table 4
(continued)

Compound	MRM	Cone voltage (V)	Collision energy (eV)	Indicative retention time (min)
RvD$_2$	375 > 175	2	22	1.26
4-HDHA	343 > 101	8	12	2.80
7-HDHA	343 > 141	6	14	2.68
10-HDHA	343 > 153	2	16	2.61
11-HDHA	343 > 193.87	2	12	2.65
13-HDHA	343 > 193.15	2	12	2.58
14-HDHA	343 > 161	12	14	2.61
17-HDHA	343 > 201	14	14	2.55
20-HDHA	343 > 241	2	12	2.48
PDX	359 > 206	18	16	1.81
MaR$_1$	359 > 177	16	16	1.82
9 OxoODE	293 > 185	14	18	2.66
13 OxoODE	293 > 113	16	20	2.58
9 HOTrE	293 > 171	20	16	2.23
13 HOTrE	293 > 195	12	16	2.26
9(10) EpOME	295 > 171	16	16	2.88
12(13) EpOME	295 > 195	2	18	2.83
Trans EKODE	309 > 209	16	10	2.30
9,10 DiHOME	313 > 201	16	20	1.98
12,13 DiHOME	313 > 183	16	20	1.91
8(9) EET-d11	330 > 155	14	12	2.96
8,9 DHET-d11	348 > 127	16	24	2.21
HXA3	335 > 273	16	12	2.29
5,15 DiHETE	335 > 115	12	12	1.82
8,15 DiHETE	335 > 155	22	16	1.76
16(17) EpDPE	343 > 233	14	12	2.87
19(20) EpDPE	343 > 285	18	12	2.79
19,20 DiHDPA	361 > 273	18	16	2.04

Table 5
MS Tune page settings for prostanoid and hydroxy fatty acid assays

Setting	Prostanoid assay	Hydroxy fatty acid assay
Capillary voltage (kV)	3.1	1.5
Source temperature (°C)	150	150
Desolvation temperature (°C)	500	600
Cone gas flow (l/h)	150	150
Desolvation gas flow (l/h)	1000	1000
LM 1 resolution	2.9	2.9
LM 2 resolution	2.8	2.8
HM 1 resolution	14.5	14.5
HM 2 resolution	14.7	14.7
Ion energy 1	1.0	1.0
Ion energy 2	0.9	0.9
MS mode entrance	1.0	1.0
MS mode exit	1.0	1.0

9. At the end of the run, program a gradient from the end conditions of the last injection up to 100% mobile phase B. Then switch to 100% shutdown solvent and flush through the column for at least 30 min, to remove any acid from the column. The column can then be removed from the UPLC and stored.

3.7 Data Analysis

1. TargetLynx™ (a part of MassLynx™ Software; Waters, Elstree, UK) can be used to construct calibration lines of each compound from the calibration standards that were run. It normalizes the peak area of each compound against the peak area of the appropriate internal standard (PGB2-*d4* for the prostanoid assay, 8(9) EET-*d11* for EETs, 8,9-DHET-*d11* for DHETs, and 12-HETE-*d8* for all other hydroxy fatty acids) to provide a "response". The response value is then plotted against the known concentration of each calibration standard and calibration lines are calculated by the least squares regression method.

2. The samples are processed in the same way, generating a "response" for each compound in each sample. The concentration corresponding to this response is then calculated and the software provides a table of the concentration of each compound in each extract (as pg/µl of extract injected).

3. Concentrations can then be multiplied by 100, to give the total amount in each sample (pg) and then normalized against sample volume (for liquid samples) or tissue wet weight or protein content (for solid tissue samples) (*see* **Note 27**).

4   Notes

1. The aqueous mobile phase should be made up fresh every 24–48 h to reduce the risk of microbial growth. We prefer not to vacuum filter the mobiles phases as this may introduce bacterial contamination.

2. We find that it is best to make mobile phase B in the original solvent bottle, since this minimizes the risk of introducing contamination, or measuring errors.

3. It is advisable to make up the seal wash a few hours in advance, since the exothermic reaction between methanol and water warms the solution and causes bubbles to form, which may elude removal by the degasser.

4. Different columns require flushing with different solvents for storage—always check the manufacturer's recommendations. Additionally, in our system we use the seal wash as the weak needle wash and mobile phase B as the strong needle wash, but again, this may vary depending on column requirements and manufacturer's recommendations.

5. Check the manufacturer's recommendations on the stability of compounds in different storage conditions.

6. Compounds should be analyzed within the linear range of the assay. If high or low concentrations are being measured and this cannot be resolved by sample dilution, extra calibration lines should be prepared at the high/low range of the assay.

7. Whenever possible, we treat cells in serum-free media, to avoid matrix effects, and remove the problem of lipid mediators that are present in the serum. Additionally, many cell culture media include lipid mediators as growth supplements—all media to be used should be examined for this and avoided if possible. Blank, unconditioned media should always be analyzed along with the conditioned media collected from cells, in order to subtract any amounts of mediators already present in the media.

8. Flat-bottomed tubes are used for the initial step since it aids the pelleting of any protein during the later centrifugation step.

9. We find that three to four drops of 0.1 M HCl are sufficient for a plasma/serum samples ≤1 ml. After adding the acid, mix the solution with a glass pipette, then take one drop and apply it to

a pH indicator strip. If the sample is not sufficiently acidic, more acid can be added and this process repeated.

10. If there is no visible pellet following centrifugation there is no need to transfer the supernatant to a clean tube.

11. Since cell culture media is well buffered it requires a stronger acid for acidification.

12. We find it useful to record the wet weight of the tissue sample, which can be used for normalization if this is preferable to protein content. The sample tube should be weighed before and after removing the tissue.

13. It is important to clean the homogenizer thoroughly between samples, using water and ethanol, to avoid cross-contamination of lipid mediators.

14. Solid tissue requires a longer incubation in the solvent than liquid samples.

15. We are not providing the protocol for protein content analysis here, since it uses a standard Bradford assay [11]. We recommend following the manufacturer's instructions.

16. It is important to keep the tube on ice as much as possible during homogenization to prevent the sample from overheating. Also, be sure to carefully remove any tissue trapped in the homogenizer to the sample tube between pulses and at the end, so that no tissue is lost.

17. Since each cartridge holds approximately 6 ml and accuracy is not important, we find it easiest to use wash bottles filled with each SPE solvent, and fill up the SPE cartridges rather than measuring out each solution for each cartridge.

18. The cartridges should be conditioned no more than 10 min before the acidified samples are ready.

19. If needed, a low vacuum can be used for highly viscous samples.

20. The hexane will require a stronger vacuum than the other solvents.

21. These washes are to remove unbound and water-soluble material (15% methanol and water, respectively), and hexane removes the water from the cartridges prior to lipid elution with methyl formate.

22. We use round-bottomed tubes for this step since they assist evaporation of the solvent.

23. Initially, a high vacuum will be required to clear the hexane, then a low vacuum is required for the methyl formate. We recommend applying a high vacuum, closing each tap once the solvent starts moving more quickly (indicating that the hexane has cleared), then switching to a low vacuum once all cartridges are clear of hexane. Finally, a high vacuum can be applied at the end to clear the last of the eluate from the cartridges.

24. We use a drying manifold with a blunt-tipped stainless-steel needle directing a flow of nitrogen into each sample tube. The stage supporting the sample tube rack can be raised and the flow of nitrogen adjusted as the solvent evaporates.

25. PDX and MaR₁ both appear in the transition of the other and cannot be separated by retention time. This means that a peak appearing in either transition could be either compound or a combination of both.

26. We run several blanks (ethanol) first to condition the system and check we have a clean baseline. Blanks should also be run between samples to allow monitoring of any carry-over. Start the run with the calibration line so you can check the quality of the LC and MS and identify any problems easily, then program calibration standards at random intervals throughout the run, to allow confirmation of retention times when processing samples. We run samples in batches of 5–6, running the prostanoid assay first, then the hydroxy fatty acid. This minimizes the amount of time that samples could evaporate following piercing of the vial in the autosampler before both assays have been run. Due to slight differences in the LC protocols between the two assays, we program two blank injections of the new assay after each switch. We run duplicate injections for each calibration standard and sample as technical repeats.

27. Some deuterated internal standards contain a small proportion of non-deuterated compound. This proportion should be estimated by the analysis of the internal standard on its own, and subtracted from the calculated amount.

Acknowledgments

We acknowledge the excellent technical support provided by Neil O'Hara, Manchester Pharmacy School, The University of Manchester.

References

1. Dennis EA, Norris PC (2015) Eicosanoid storm in infection and inflammation. Nat Rev Immunol 15(8)511–523

2. Ricciotti E, FitzGerald GA (2011) Prostaglandins and inflammation. Arterioscler Thromb Vasc Biol 31(5)986–1000

3. Massey KA, Nicolaou A (2013) Lipidomics of oxidized polyunsaturated fatty acids. Free Radic Biol Med 59:45–55

4. Bannenberg G, Serhan CN (2010) Specialized pro-resolving lipid mediators in the inflammatory response: an update. Biochim Biophys Acta 1801(12)1260–1273

5. Deems R, Buczynski MW, Bowers-Gentry R, Harkewicz R, Dennis EA (2007) Detection and quantitation of eicosanoids via high performance liquid chromatography-electrospray ionization-mass spectrometry. Methods Enzymol 432:59–82

6. Puppolo M, Varma D, Jansen SA (2014) A review of analytical methods for eicosanoids in brain tissue. J Chromatogr B 964:50–64

7. Astarita G, Kendall AC, Dennis EA, Nicolaou A (2015) Targeted lipidomic strategies for oxygenated metabolites of polyunsaturated fatty acids. Biochim Biophys Acta 1851(4)456–468

8. Murphy RC, Barkley RM, Zemski Berry K, Hankin J et al (2005) Electrospray ionization and tandem mass spectrometry of eicosanoids. Anal Biochem 346(1)1–42

9. Masoodi M, Nicolaou A (2006) Lipidomic analysis of twenty-seven prostanoids and isoprostanes by liquid chromatography/electrospray tandem mass spectrometry. Rapid Commun Mass Spectrom 20(20)3023–3029

10. Masoodi M, Mir AA, Petasis NA, Serhan CN, Nicolaou A (2008) Simultaneous lipidomic analysis of three families of bioactive lipid mediators leukotrienes, resolvins, protectins and related hydroxy-fatty acids by liquid chromatography/electrospray ionisation tandem mass spectrometry. Rapid Commun Mass Spectrom 22(2)75–83

11. Bradford MM (1976) A rapid and sensitive method for the quantitation of microgram quantities of protein utilizing the principle of protein-dye binding. Anal Biochem 72:248–254

Chapter 16

Derivatization of Fatty Aldehydes and Ketones: Girard's Reagent T (GRT)

Paul Wood

Abstract

In lipid metabolism and during oxidative stress, a number of fatty aldehydes are generated. Detection and quantitation of these reactive aldehydes is optimally achieved by trapping them with a derivatization reagent. In this chapter, we present Girard's Reagent T (GRT) that is a robust agent for derivatizing aldehydes and ketones and provides a derivative that is highly charged for accurate ESI quantitation. GRT is a quaternary amine with a reactive hydrazine group that derivatizes aldehydes and ketones to form hydrazones:

$$R - CHO + NH_2 - NH - CO - CH_2 - N(CH_3)_3 \rightarrow R - C = N - NH - CO - CH_2 - N(CH_3)_3$$

The aldehydes or ketones are augmented in mass by 113.09528 after derivatization with GRT and the $[M+H]^+$ ions of these derivatives are monitored in ESI. In this chapter, we describe the use of GRT for quantitating fatty aldehydes (Johnson, Rapid Commun Mass Spectrom 21:2926–2932, 2007) ketosterols (Johnson, Rapid Commun Mass Spectrom 21:2926–2932, 2007; Johnson, Rapid Commun Mass Spectrom 19:193–200, 2005; Crick et al., Biochem Biophys Res Commun 446:756–761, 2014), and lipid oxidation products like formyl lysophosphatidylcholine (Fuchs et al., Lipids 42:991–998, 2007).

Key words Fatty aldehydes, Girard's reagent T, Amniotic fluid

1 Introduction

There are a number of reagents that are useful for derivatizing, and thereby stabilizing, reactive aldehydes and ketones. In the case of mass spectrometric analyses utilizing ESI, it is important to introduce a charged moiety to improve sensitivity in quantitative experiments. GRT is such a robust and useful reagent [1–3].

2 Materials and Equipment

1. **Reagent**: 20 mg. Girard's reagent T per mL of methanol.

Paul Wood (ed.), *Lipidomics*, Neuromethods, vol. 125,
DOI 10.1007/978-1-4939-6946-3_16, © Springer Science+Business Media LLC 2017

2. **Infusion solvent**: 80 mL of 2 propanol + 40 mL methanol + 20 mL chloroform + 0.5 mL water containing 164 mg ammonium acetate.

3. **Stable isotope internal standards**: [$^2H_5$]hexadecanal and [$^2H_5$]hexadecenal are purchased from Avanti Polar Lipids.

4. **Mass spectrometer**: Q Exactive (Thermo Fisher).

3   Methods

3.1 Extraction (MTBE)

1. For biofluids, 100 μL are mixed with 1 mL of methanol containing the internal standard mixture and 1 mL of water. In the case of tissues and cells, they are sonicated in the methanol–water solution.

2. Next 2 mL of t-butylmethyether (MTBE) are added and the tubes capped. They are next shaken vigorously at room temperature for 30 min using a Fisher Multitube Vortexer.

3. The tubes are then centrifuged at $3000 \times g$ for 15 min at room temperature.

4. Next 1 mL of the upper organic lipid layer is transferred to a screw-top 1.5 mL microfuge tube and the samples dried for 3.5 h in an Eppendorf Vacufuge.

5. The samples are redissolved in 150 μL of the Infusion Solvent and the tubes centrifuged at $30,000 \times g$ for 5 min in an Eppendorf Microfuge to precipitate any particulates.

3.2 Extraction (AM)

1. For more polar metabolites we utilize acetonitrile/methanol (AM; 8:1) as the extraction solvent. We utilize 1 mL of AM per 0.1 mL of plasma or 30–40 mg of tissue. Biofluids are mixed with the solvent while tissues are sonicated.

2. Next the samples are centrifuged at $30,000 \times g$ for 30 min at 4 °C.

3. For abundant aldehydes, 0.1 mL of the supernatant can be reacted directly with an equal volume of the reagent or can be dried by vacuum centrifugation prior to derivatization.

3.3 Derivatization

1. To the dried extract in 1.5 mL screw-top microvials, add 0.1 mL of reagent and 20 μL of acetic acid and heat at 70 °C for 30 min with shaking (Eppendorf Mutitube Heater).

2. The samples are then dried by vacuum centrifugation prior to dissolution in the infusion solvent.

3.4 High-Resolution Mass Spectrometric Analysis

Samples are infused at 5 μL/min and analyzed via positive ion ESI (<3 ppm mass error).

3.5 Tandem Mass Spectrometry

The dominant ions observed with MS^2 are $[M-59.0735]^+$ and $[M-87.0684]^+$ which correspond to $[M-N(CH_3)_3]^+$ and $[M-CO-N(CH_3)_3]^+$ losses from the GRT, respectively [4, 5].

3.6 Mass Spectrometric Data Reduction

For the high-resolution mass spectrometric data, the top 1000 masses, for a 200–300 amu window, and their associated ion intensities are transferred to an Excel spreadsheet. Within the spreadsheet is a list of aldehydes, their exact masses (five decimals), and the scanned ions for the GRT derivatives which are searched within the data table. If the calculated ppm mass error is determined to be ≤3 for the extracted ion, then the ratio of the ion intensity to that of the assigned internal standard is calculated and added to the spreadsheet. An example of data obtained with this method is presented in Table 1.

Table 1
GRT-reactive lipids extracted from 100 μL equine amniotic fluid

Lipid	Exact	Derivative [M+H]⁺	Observed	ppm
Int. Std.				
[⁵H₂]Hexadecanal	245.2767	359.3793	359.3784	0.83
Aldehydes				
Acrolein	56.0262	170.1288	170.1287	0.53
4-HNE	156.1150	270.2176	270.2171	0.46
Hexadecanal	240.2453	354.3479	354.3475	0.37
Octadecanal	268.2766	382.3792	382.3790	0.21
Neurosteroids				
Pregnenolone	316.2402	430.3428	430.3426	0.20
Allopregnanolone	318.2559	432.3584	432.3583	0.19
Formyl Lipids				
Formyl LPC 14:1	493.2804	607.3830	607.3817	1.33
Formyl LPC 16:1	521.3118	635.4143	635.4127	1.64
Formyl LPC 18:6	539.2648	653.3674	653.3659	1.47
Formyl LPE 16:1	479.2648	593.3674	593.3660	1.40
Formyl LPE 18:1	507.2961	621.3987	621.3970	1.68
Formyl LPE 20:6	525.2492	639.3517	639.3511	0.61
Formyl LPE 22:5	555.2961	669.3987	669.3972	1.45

4-HNE, 4-hydroxynonenal; LPC, lysophosphatidylcholine; LPE, lysophosphatidylethanolamine; Ppm, parts per million mass error

4   Conclusion

GRT is a robust reagent that generates stable and sensitive derivatives of aldehydes and keto lipids.

Acknowledgments

This work was funded by Lincoln Memorial University.

References

1. Johnson DW (2007) A modified Girard derivatizing reagent for universal profiling and trace analysis of aldehydes and ketones by electrospray ionization tandem mass spectrometry. Rapid Commun Mass Spectrom 21:2926–2932
2. Johnson DW (2005) Ketosteroid profiling using Girard T derivatives and electrospray ionization tandem mass spectrometry: direct plasma analysis of androstenedione, 17-hydroxyprogesterone and cortisol. Rapid Commun Mass Spectrom 19:193–200
3. Crick PJ, Aponte J, Bentley TW, Matthews I, Wang Y, Griffiths WJ (2014) Evaluation of novel derivatisation reagents for the analysis of oxysterols. Biochem Biophys Res Commun 446:756–761
4. Fuchs B, Müller K, Göritz F, Blottner S, Schiller J (2007) Characteristic oxidation products of choline plasmalogens are detectable in cattle and roe deer spermatozoa by MALDI-TOF mass spectrometry. Lipids 42:991–998
5. Griffiths WJ, Alvelius G, Liu S, Sjövall J (2004) High-energy collision-induced dissociation of oxosteroids derivatised to Girard hydrazones. Eur J Mass Spectrom (Chichester) 10:63–88

Chapter 17

Derivatization of Lipid Amines: Fluorenylmethyloxycarbonyl (FMOC)

Paul Wood

Abstract

Analysis of lipid remodeling can be achieved by monitoring the lyso metabolites of ethanolamine glycerophospholipids. However, mass spectrometric analyses cannot distinguish between alkenyl and alkyl lyso-glycerophosphoethanolamines without chromatography. In this chapter, we describe the use of fluorenylmethyloxycarbonyl chloride (FMOC) to distinguish these isobars by direct infusion and MS^2. FMOC is a useful reagent for derivatizing the free amine group of lipid ethanolamines (Han et al., J Lipid Res 46:1548–1560, 2005):

$$R - NH_2 + Cl - CO - O - CH_2 - \text{Fluren} \rightarrow R - NH - CO - O - CH_2 - \text{Fluren}$$

The amine-containing lipids are augmented in mass by 222.06808 after derivatization with FMOC and the [M-H]$^-$ ions of these derivatives are monitored in ESI. In this chapter, we describe the use of FMOC for quantitating ethanolamine plasmalogens, lysophosphatidylethanolamines (LPE), alkenyl-LPEs, and alkyl-LPEs.

Key words Lipid amines, FMOC

1 Introduction

Lipid remodeling at sn-2 of glycerophospholipids is a dynamic process [1, 2].

2 Materials and Equipment

1. **Reagent**: 2 mg. FMOC.Cl per mL of chloroform.

2. **Catalyst**: 1% ammonium acetate (Note: in several published studies of lipid derivatization with FMOC, a catalyst was not included. Without the catalyst, derivatization is only 50% complete while with the catalyst >99% derivatization is achieved).

Paul Wood (ed.), *Lipidomics*, Neuromethods, vol. 125,
DOI 10.1007/978-1-4939-6946-3_17, © Springer Science+Business Media LLC 2017

3. **Infusion Solvent**: 80 mL of 2 propanol + 40 mL metha-nol + 20 mL chloroform + 0.5 mL water containing 164 mg ammonium acetate.

4. **Internal Standard**: Lyso PE 13:0 is purchased from Avanti Polar Lipids.

5. **Mass Spectrometer**: Q Exactive (Thermo Fisher).

3   Methods

3.1   Extraction

1. For biofluids, 100 μL is mixed with 1 mL of methanol contain-ing the internal standard mixture and 1 mL of water. In the case of tissues and cells, they are sonicated in the methanol–water solution.

2. Next 2 mL of t-butylmethyether (MTBE) is added and the tubes capped. They are next shaken vigorously at room tem-perature for 30 min using a Fisher Multitube Vortexer.

3. The tubes are then centrifuged at 3000 × g for 15 min at room temperature.

4. Next 1 mL of the upper organic lipid layer is transferred to a screw-top 1.5 mL microfuge tube and the samples dried for 3.5 h in an Eppendorf Vacufuge.

5. The samples are redissolved in 150 μL of the Infusion Solvent and the tubes centrifuged at 30,000 × g for 5 min in an Eppendorf Microfuge to precipitate any particulates.

3.2   Derivatization

1. To the dried extract in 1.5 mL screw-top microvials, add 0.1 mL of reagent and 20 μL of 1% ammonium acetate prior to heating at 40 °C for 15 min with shaking (Eppendorf Multitube Heater). The base catalysis is essential for complete conversion of ethanolamine lipids to their FMOC derivatives.

2. The samples are then dried by vacuum centrifugation prior to dissolution in the infusion solvent.

3.3   High-Resolution Mass Spectrometric Analysis

Samples are infused at 5 μL/min and analyzed via negative ion ESI (<3 ppm mass error).

3.4   Tandem Mass Spectrometry

In tandem analyses, the dominant MS^2 ion with:

1. LPEs and ethanolamine plasmalogens is the anion of the acyl-fatty acid substituent.

2. Alkenyl LPEs is 196.03748 (glycerophosphoethanolamine—H_2O), resulting from the loss of FMOC and the alkenyl substituent.

3. Alkyl LPEs is 140.0113 (phosphoethanolamine).

3.5 Mass Spectrometric Data Reduction

For the high-resolution mass spectrometric data, the top 1000 masses, for 200–300 amu windows, and their associated ion intensities are electronically transferred to an Excel spreadsheet. Within the spreadsheet is a list of FMOC derivatives, their exact masses (five decimals), and the scanned ions for the derivatives which are searched within the data table. If the calculated ppm mass error is determined to be ≤3 for the extracted ion, then the ratio of the ion intensity to that of the assigned internal standard is calculated and added to the spreadsheet (Tables 1, 2, and 3).

Table 1
MS² of the FMOC derivatives of lysophosphatidylethanolamines (LPE)

LPE	Exact	FMOC [M-H]⁻	Product @ 25 [M-H]⁻
LPE 13:0 (Int Std)	411.2386	632.2994	213.1860
LPE 16:0	453.2855	674.3463	255.2329
LPE 16:1	451.2699	672.3307	253.2173
LPE 18:0	481.3168	702.3776	283.2642
LPE 18:1	479.3012	700.3620	281.2486
LPE 18:2	477.2855	698.3463	279.2329
LPE 18:3	475.2699	696.3307	277.2173
LPE 20:4	501.2855	722.3463	303.2329
LPE 22:6	525.2855	746.3463	327.2329

MS² at 10 results in the loss of FMOC (−222.06808) while at 25 the product is the acyl fatty acid substituent

Table 2
MS² of the FMOC derivatives of alkenyl-lysophosphatidylethanolamines (LPEp)

LPEp	Exact	FMOC [M-H]⁻	Product @ 25 [M-H]⁻
LPEp 16:0	437.2906	658.3514	196.0374
LPEp 16:1	435.2749	656.3357	196.0374
LPEp 18:0	465.3219	686.3827	196.0374
LPEp 18:1	463.3062	684.3670	196.0374
LPEp 18:2	461.2905	682.3513	196.0374
LPEp 18:3	459.2748	680.3356	196.0374
LPEp 20:4	485.2904	706.3512	196.0374
LPEp 22:6	517.3532	738.4140	196.0374

MS² at 10 results in the loss of FMOC (−222.06808) while at 25 the product is the loss of FMOC and the plasmenyl substituent
Note: in the case of the FMOC derivatives of alkyl-LPE, MS² at 10 results in the loss of FMOC (−222.06808) while at 25 the product is ethanolamine (140.0113)

Table 3
MS^2 of the FMOC derivatives of alkenyl-acyl glycerophosphoethanolamines (ethanolamine plasmalogens; PlsE)

PlsE	Exact	FMOC [M-H]$^-$	Product @ 25 [M-H]$^-$
PlsE 16:0/16:0	675.5203	896.5811	255.2329
PlsE 16:0/18:0	703.5516	924.6124	283.2642
PlsE 16:0/18:1	701.5359	922.5967	281.2486
PlsE 16:0/18:2	699.5203	920.5811	279.2329
PlsE 16:0/18:3	697.5046	918.5654	277.2173
Pls E 16:0/20:4	723.5203	944.5811	303.2329
PlsE 16:0/22:6	747.5203	968.5811	327.2329

MS^2 at 10 results in the loss of FMOC (-222.06808) while at 25 the product is the sn2 fatty acyl substituent

4   Conclusion

FMOC is a robust reagent that generates stable and sensitive derivatives of amine-containing lipids allowing the determination of both alkyl and alkenyl lysoglycerophospho-ethanolamines via direct infusion and MS^2 analyses.

Acknowledgments

This work was funded by Lincoln Memorial University.

References

1. Han X, Yang K, Cheng H, Fikes KN, Gross RW (2005) Shotgun lipidomics of phosphoethanolamine-containing lipids in biological samples after one-step in situ derivatization. J Lipid Res 46:1548–1560

2. Wood PL (2012) Lipidomics of Alzheimer's disease: current status. Alzheimers Res Ther 4:5

Chapter 18

High-Resolution Mass Spectrometry of Glycerophospholipid Oxidation Products

Paul Wood

Abstract

Oxidation of lipids occurs in vivo with oxidative stress and during storage of biological samples, even at −80 °C. In this chapter, we demonstrate that the oxidation products of phosphatidylcholines include an aldehyde (sn-2 unsaturated fatty acyl substituent), an acid (sn-2 unsaturated fatty acyl substituent), the associated lysophosphatidylcholine, and a hydroperoxide (Fuchs et al., Lipids 42:991–998, 2007; Hui et al., Anal Bioanal Chem 403:1831–1840, 2012).

Key words Lipid oxidation, Hydroperoxides, Lipid acids, Lyso lipids, Lipid aldehydes

1 Introduction

By utilization of diacyl and alkenyl-acyl glycerophospholipids as model lipids, we determined which oxidation products are generated after oxidation for 24 h at room temperature [1, 2].

2 Materials and Equipment

1. **Lipids**: phosphatidylcholine 18:0/22:6, choline plasmalogen 18:0/22:6, phosphatidylethanolamine 18:0/22:6, and ethanolamine plasmalogen 18:0/22:6 (Avanti Polar Lipids).

2. **Infusion Solvent**: 80 mL of 2 propanol + 40 mL methanol + 20 mL chloroform + 0.5 mL water containing 164 mg ammonium acetate.

3. **Mass Spectrometer**: Q Exactive (Thermo Fisher).

Paul Wood (ed.), *Lipidomics*, Neuromethods, vol. 125,
DOI 10.1007/978-1-4939-6946-3_18, © Springer Science+Business Media LLC 2017

3 Methods

3.1 Oxidation

1. 1 nanomole of lipid was dried in a 1.5 mL microtube and left open to the air at room temperature overnight.
2. Next the sample was dissolved in infusion solvent.
3. The tubes are then centrifuged at $30,000 \times g$ for 5 min in an Eppendorf Microfuge to precipitate any particulates.
4. In the case of ethanolamine phospholipids, the lyso lipids as well as the parent lipids can be derivatized with FMOC (Chap. 17) and monitored via direct infusion.

3.2 High-Resolution Mass Spectrometric Analysis

Samples are infused at 5 µL/min and analyzed via positive ion ESI (<3 ppm mass error).

3.3 Mass Spectrometric Data Reduction

The extracted ions for the parent lipid, the aldehyde, the acid, the hydroperoxide, and the lyso lipid were monitored. As Presented in Tables 1 and 2, the relative ion intensities were different for diacyl vs. alkenyl-acyl glycerophospholipids. Table 3 presents the FMOC derivatives of the lysophosphatidylethanolamines generated during lipid oxidation, while Table 4 lists the FMOC derivatives of the parent ethanolamine plasmalogens.

4 Conclusion

Monitoring the aldehyde and acid metabolites of glycerophospholipids is a robust method to monitor lipid oxidation. The hydroperoxide is a sensitive biomarker of plasmalogen oxidation, while acid metabolites are useful for monitoing oxidation of diacyl glycerophospholipids. The relative contributions of in vivo vs. sample storage conditions complicate interpretation of the data without well-matched controls.

Acknowledgments

This work was funded by Lincoln Memorial University.

Table 1
Oxidation products of phosphatidylcholine 18:0/22:6

PtdCh 18:0/22:6
$C_{48}H_{84}NO_8P$; 833.5935
$[M+H]^+ = 834.6007$ (0.83 ppm)

  LPC 18:0 $C_{26}H_{54}NO_7P$; 523.3638 $[M+H]^+ = 524.3710$ (0.32 ppm) Relative intensity = 11.5	$H_{35}C_{17}\text{-CO-O-CH}_2$ | $OHC\text{-CH}_2\text{-CH}_2\text{-CO-O-CH}$ | $CH2\text{-O-PC}$ PtdCh 18:0/22:6 Aldehyde $C_{30}H_{58}NO_9P$; 607.3849 $[M+H]^+ = 608.3921$ (0.10 ppm) Relative intensity = 100
PtdCH 18:0/22:6 hydroperoxide $C_{48}H_{84}NO_{10}P$; 865.5181 $[M+H]^+ = 866.5906$ (0.92 ppm) Relative intensity = 7.8	$H_{35}C_{17}\text{-CO-O-CH}_2$ | $HOOC\text{-CH}_2\text{-CH}_2\text{-CO-O-CH}$ | $CH2\text{-O-PC}$ PtdCh 18:0/22:6 Acid $C_{30}H_{58}NO_{10}P$; 623.3798 $[M+H]^+ = 624.3871$ (0.48 ppm) Relative intensity = 66.4

Table 2
Comparison of the oxidation products of phosphatidylcholine 18:0/22:6 (PtdC), choline plasmalogen 18:0/22:6 (PlsC), phosphatidylethanolamine 18:0/22:6 (PtdE), and ethanolamine plasmalogen 18:0/22:6(PlsE)

Lipid	Aldehyde	Acid	Lyso	Hydroperoxide
PtdC 40:6	100	66.4	11.5	7.8
PlsC 40:6	29.5	9.6	1.0	100
PtdE 40:6	13.1	89.5	100	33.5
PlsE 40:6	5.8	46.9	64.6	100

The listed values are relative intensities

Table 3
MS^2 of the FMOC derivatives of alkenyl-lysophosphatidylethanolamines (LPEp). MS^2 at 10 results in the loss of FMOC (−222.06808) while at 25 the product is the loss of FMOC and the plasmenyl substituent

LPEp	Exact	FMOC [M−H]$^-$	Product @ 25 [M−H]$^-$
LPEp 16:0	437.2906	658.3514	196.0380
LPEp 16:1	435.2749	656.3357	196.0380
LPEp 18:0	465.3219	686.3827	196.0380
LPEp 18:1	463.3062	684.3670	196.0380
LPEp 18:2	461.2905	682.3513	196.0380
LPEp 18:3	459.2748	680.3356	196.0380
LPEp 20:4	485.2904	706.3512	196.0380
LPEp 22:6	517.3532	738.4140	196.0380

Note: in the case of the FMOC derivatives of alkyl-LPE, MS^2 at 10 results in the loss of FMOC (−222.06808) while at 25 the product is $C_5H_{12}NO_5P$ (anion of 196.0380)

Table 4
MS^2 of the FMOC derivatives of alkenyl-acyl glycerophosphoethanolamines (ethanolamine plasmalogens; PlsE)

PlsE	Exact	FMOC [M-H]$^-$	Product @ 25 [M-H]$^-$
PlsE 16:0/16:0	675.5203	896.5811	255.2329
PlsE 16:0/18:0	703.5516	924.6124	283.2642
PlsE 16:0/18:1	701.5359	922.5967	281.2486
PlsE 16:0/18:2	699.5203	920.5811	279.2329
PlsE 16:0/18:3	697.5046	918.5654	277.2173
Pls E 16:0/20:4	723.5203	944.5811	303.2329
PlsE 16:0/22:6	747.5203	968.5811	327.2329

MS^2 at 10 results in the loss of FMOC (−222.06808) while at 25 the product is the sn2 fatty acyl substituent

References

1. Fuchs B, Müller K, Göritz F, Blottner S, Schiller J (2007) Characteristic oxidation products of choline plasmalogens are detectable in cattle and roe deer spermatozoa by MALDI-TOF mass spectrometry. Lipids 42:991–998

2. Hui SP, Taguchi Y, Takeda S, Ohkawa F, Sakurai T, Yamaki S, Jin S, Fuda H, Kurosawa T, Chiba H (2012) Quantitative determination of phosphatidylcholine hydroperoxides during copper oxidation of LDL and HDL by liquid chromatography/mass spectrometry. Anal Bioanal Chem 403:1831–1840

INDEX

Paul Wood (ed.), *Lipidomics*, Neuromethods, vol. 125,
DOI 10.1007/978-1-4939-6946-3, © Springer Science+Business Media LLC 2017

Printed in the United States
By Bookmasters